MISSION TO MARS

ALSO BY MICHAEL COLLINS

Carrying the Fire: An Astronaut's Journeys
Flying to the Moon and Other Strange Places
Liftoff: The Story of America's Adventure in Space

MISSION TO MARS

MICHAEL COLLINS

GROVE WEIDENFELD
New York

Published by Grove Weidenfeld
A division of Grove Press, Inc.
841 Broadway
New York, NY 10003-4793

Published in Canada by General Publishing Company, Ltd.

Illustrations were provided courtesy of: MARK SEIDLER: charted route. ROY ANDERSEN: module diagram; problems of daily stress; collage of colony activities. PIERRE MION: round-trip module departing Earth; ships entering Martian orbit. U.S. GEOLOGICAL SURVEY, FLAGSTAFF, ARIZONA: Mars/Valles Marineris; eastern Mangala Vallis. NASA: Chryse Planitia; *Viking 1* self-portrait; valley in Antarctica; ships of exploration.

The Mars Declaration is reproduced courtesy of The Planetary Society.

Library of Congress Cataloging-in-Publication Data

Collins, Michael, 1930–
Mission to Mars / Michael Collins. — 1st ed.
p. cm.
ISBN 0-8021-1160-2
1. Space flight to Mars. I. Title.
TL799.M3C65 1990
629.45′53—dc20
90-37632
CIP

Manufactured in the United States of America

Printed on acid-free paper

First Edition

Designed by Irving Perkins Associates

10 9 8 7 6 5 4 3 2 1

To those who will make the trip.

When you do, I'll be like the kid
in a small town, listening to the
big diesels growl through at midnight,
wanting to climb on board.

ACKNOWLEDGMENTS

I found a lot of people interested in Mars, even to the extent of helping me with this book. Their contributions were varied and complex; rather than attempting to separate them by category, I list them in alphabetical order. Each on the list is important, and I thank them all.

Aaron Asher
Silvio Bedini
Susan Bonner
Tom Canby
Mark Craig
Louis Friedman
Howard Golden
Gregor Hall
Fred Jordan
Gene Marianetti
Barbara McConnell
John Niehoff
Howard Paine
Terry Pietroski
Roger Ressmeyer
Jon Schneeberger
Ed Sedarbaum
Abigail Tipton
Doug Ward
Jack Whitelaw
Ruth Winston

CONTENTS

FOREWORD

A lot has happened in the months since I finished writing this book. It is now clear that George Bush is willing to put money and political clout behind his gung ho space speech of July 20, 1989. For fiscal year 1991 he has asked Congress for $15.2 billion for NASA, a 24 percent annual increase. It looks as if he will get less, around $14 billion, but that still makes NASA the fastest-growing major government agency. In these days of roaring deficits and Gramm-Rudman restrictions, that fact itself is extraordinary.

On the negative side, Congress finds Moon-Mars planning the least urgent of NASA's tasks, and the House of Representatives has cut nearly all Moon-Mars research funds. This money may be restored by the Senate or obtained from the military. The president has vowed to fight the delay, saying that "had Columbus waited until all the problems of his time were solved, the timbers of the *Santa María* would be rotting on the Spanish coast to this very day." The president has also picked a timetable for Mars, stating, "I believe that before Apollo celebrates the fiftieth anniversary of its landing on the Moon, the American flag should be planted on Mars." He did not mention foreign participation in this 2019 mission, but the White House has announced that the United States will seek an "exploratory dialogue" with European nations, Canada, Japan, the Soviet Union, and other countries. The Soviets are still talking about joint U.S.-U.S.S.R. missions that could land on Mars between 2005 and 2010.

The price tag seems to have gone up, and NASA is talking about needing as much as $500 billion over the next thirty years. A major fraction of that, in my opinion, is caused by requiring a Moon base first. I have argued in this book that it makes better sense to bypass the Moon entirely and proceed directly to Mars. We could certainly get there long before 2019. I worry that a Moon base will develop into a competitor for Mars funds, delaying indefinitely the exploration of a much more interesting and useful place. As far as the price tag goes, many credible experts claim $500 billion is vastly inflated. Above all, NASA wants plenty of money for space station *Freedom,* and approaches its tasks in a serial fashion: *Freedom,* then the Moon, and Mars a distant third.

The National Space Council, chaired by Vice President Dan Quayle, has put pressure on NASA to abandon its business-as-usual thinking and consider innovative ways to reduce costs and accelerate schedules. NASA has responded with an "Outreach Program," asking technical societies, other agencies, and even individuals to submit ideas. From these will emerge, supposedly by early 1991, "reference architecture, a technology assessment, and possible early milestones."

But in the meantime NASA has more immediate problems. Its shuttle is grounded by fuel leaks, and the long-awaited Hubble telescope is finally in orbit, but performing poorly. An enthusiastic president, a hard-luck space agency, an ambivalent Congress: to paraphrase Charles Dickens, it's the best of times, and the worst of times.

Michael Collins
July 4, 1990

MISSION TO MARS

CHAPTER

I

THROUGH A TELESCOPE

THEIR BEAUTIFUL blue and white Earth was gone now, its remnant a mere pinpoint of light in their window, and no brighter than Venus. Each of the crew thought about "home" differently. Their families, of course, were on all their minds, and their old neighborhoods to some extent, and their individual recollections of the sun-drenched grandeur of Earth orbit. But when they gathered around their wardroom table and looked out their only large porthole, they discovered that there was an overriding "it" that attracted them, that pulled on their emotions, that transcended even the memories of their children's faces. The planet itself, no particular spot on it, acted as a powerful magnet, but like a magnet it lost some power with distance as it faded far away. They sensed its diminished presence as a real loss, a void filled by sadness but also by an increased resolve to persevere without their celestial anchor.

Mars was not yet close enough to make its presence really felt, although week by week it grew redder and plumper. The crew enjoyed discussing their impressions with one another and with Mission Control, although with a ten-minute time delay conversations with Earth were increasingly stilted. Instead of just talking, both the crew and the ground controllers tended to make organized speeches. The formality of it was hard to avoid, despite their best efforts. Among the crew, however, camaraderie was increasing: each day reinforced the realization that they were truly alone, on their own, and would succeed or fail irrespective of the stream of advice and good will that linked them to Earth. They were proud of themselves and the fact that—so far at least—they were all strong and well, physically and mentally. Even their plants seemed to be thriving.

The decision to launch the mission had been long and tortuous, its roots as old as the telescope. As far back as the seventeenth century astronomers had marveled at the fuzzy orange ball in their eyepieces, and lavished on it their most fanciful dreams. It was natural for them to imagine life there, and as telescopes became more powerful, the "evidence" grew. For example, in the late nineteenth century, the Italian Giovanni Schiaparelli reported thin, nearly straight lines etched in the Martian surface. He called these grooves *canali*, but when translated into English "canals" suggested constructed artifacts rather than the geologic formations Schiaparelli had described. Canals must be intended, thought many, for the irrigation of desert cities—clearly, intelligent beings inhabited Earth's sister planet. Probably the most influential proponent of this view was

Percival Lowell, an astronomer and writer born in Boston in 1855. Lowell devoted his life to studying and describing Mars. He concluded not only that Mars was inhabited but that the Martians were an older and more advanced civilization, forced by a shrinking water supply to develop superior technologies. Lowell was widely read and believed.

At the turn of the century H. G. Wells published *The War of the Worlds*, describing a fierce attack on Earth by Martians. Edgar Rice Burroughs, creator of Tarzan, wrote swashbuckling Martian tales long before he switched to jungle adventures. In recent years the most widely read and admired book about Mars is probably Ray Bradbury's *The Martian Chronicles*, a collection of poignant, lyrical stories about a dying Martian civilization.

In 1938 CBS radio aired a version of *The War of the Worlds*. Narrated by Orson Welles, bulletins of a purported Martian invasion of New Jersey were interspersed with dance music. People who tuned in after the initial disclaimer were convinced that an actual invasion was taking place, by tentacled creatures with V-shaped mouths dripping saliva from rimless lips. Panic ensued, thousands fled their homes, and the nationwide publicity put the Red Planet on everyone's front page.

Perhaps because of its proximity to Earth, Mars has captured our imagination in ways that the other planets, such as the gaseous giants Jupiter and Saturn, have not—despite spectacular close-up photographs of them in recent years. Unlike the others, Mars seems friendly, accessible, even habitable. As author James Michener has put it: "Mars has played a special role in our lives, because of the literary and philosophical speculations that have centered upon it. I have always known Mars."

I grew up not only knowing the place but wanting to go there. I joke that *Apollo 11* took me to the wrong planet, to the Moon instead of Mars, but it is true that a close-up look at the Moon has served only to whet my Martian appetite. It is ironic, I think, that after the flight of *Apollo 11* in 1969, Neil Armstrong, Buzz Aldrin, and I were awarded the Guzman Medal by the French Academy of Sciences. The medal had been waiting to be awarded since 1889. It went to the first persons "to find the means of communicating with a heavenly body—Mars excluded." The exception of Mars was stipulated by the medal's founder, Anna Emile Guzman, "because that planet appears to be sufficiently well known." I don't know how Guzman managed to fill her Martian scrapbook, but in the century since 1889 we have reached out to Mars in many ways, and today we can describe conditions there with some clarity. As usual with scientific advances, however, each answered question seems to spawn several new puzzles.

In May 1961 President John F. Kennedy decided that the United States should go to the Moon by the end of the decade, and it was done with five months to spare. But how about Mars? What would be involved in such a stupendous voyage? When and how might we reasonably expect to go there? What might we find? Is Kennedy's Project Apollo a model to be emulated or discarded? To begin our voyage it might be good to consider where we Earthlings are and where and what Mars is.

CHAPTER

II

MARINERS AND VIKINGS

OUR SUN is a very ordinary star buried in one of the spiral arms of the Milky Way galaxy. Nine planets orbit the Sun; Mercury and Venus are closer to it than Earth, the others farther away. Earth is about 92 million miles from the Sun; Mars is next in line, about 140 million miles out. Beyond Mars lies a belt of asteroids, customarily considered the dividing line between the four inner planets and the outer ones—Jupiter, Saturn, Uranus, Neptune, and Pluto. The orbit of Pluto is cocked at an angle to that of all the other planets, which revolve around the Sun in what is called the Plane of the Ecliptic. All the planets move in the same direction, counterclockwise as viewed from above Earth's northern hemisphere. It takes the Earth twenty-four hours to turn once on its axis (a day) and 365¼ days to make one trip around the Sun (a year). Our calendar acknowledges the messiness of that quarter day by declaring every fourth year "leap year," and February 29 is slipped in to make things right again.

The seasons occur because the Earth's axis is not perpendicular to the Plane of the Ecliptic but is tilted 23½ degrees to it, so that sometimes one hemisphere is pointed more directly at the Sun than is the other. The hemisphere facing the Sun more directly experiences summer while it's winter in the other half. Maps note this fact with a line at 23½ degrees north latitude called the Tropic of Cancer and a similar one south, the Tropic of Capricorn.

Mars, being farther away from the Sun, has a bigger orbit than Earth's, and it takes longer—687 days—to complete one turn around the Sun, making its year nearly twice as long as that of Earth. Its day, however, is remarkably similar, only thirty-seven minutes longer than our twenty-four hours. Its seasons are also similar, because Mars is tilted 25 degrees to the Plane of the Ecliptic at an angle comparable to 23½ degrees.

Instead of one moon, Mars has two—Phobos and Deimos. Compared to ours they are tiny and closer in. Phobos, about seventeen miles long, takes less than eight hours for one orbit, while Deimos, about half the size of Phobos, circles once every thirty hours. Thus the concept of a lunar month doesn't mean much on Mars. Both Phobos and Deimos look somewhat like blackened potatoes, with surfaces pockmarked by millions of craters. They are possibly of material similar to that found in stony meteorites. Phobos is frequently suggested as a good place for people to land and inspect Mars at close quarters. A Phobos mission would not require the extra fuel needed to fight Mars's gravity all the way down to its surface on landing and then back up again.

Mars itself is slightly more than half the size of Earth and has a gravitational field only .38 as strong. In other words, a hundred-pound person would weigh thirty-eight pounds on Mars, a pretty good weight for jumping around but still twice as heavy as on our own moon. Being half as many miles again farther away from the Sun than is Earth, Mars should be a cold place—and it is. At its very hottest, midsummer near the equator, the surface may reach 70 degrees Fahrenheit, but generally temperatures are in the cruel subzero range, and in winter the poles may reach 200 degrees below.

Before unprotected humans on Mars could freeze to death, however, they would succumb to asphyxiation. The Martian atmosphere is extremely thin and is 95 percent carbon dioxide, with small amounts of nitrogen and argon and a trace of oxygen, carbon monoxide, and water vapor. Its pressure is so low—the equivalent of the Earth's atmosphere at an altitude of 100,000 feet—that human blood would boil. To prevent this, human hikers on Mars will have to wear pressure suits similar to those of the Apollo astronauts. But unlike on the surface of the Moon, which has no atmosphere whatsoever, Martian travelers will have to contend with howling winds that can produce dust storms lasting for months and filling the sky with a pale pink or orange haze.

Despite these extreme conditions, Mars is better suited for human habitation than any other planet in the Solar System. If we are to establish ourselves as a multiplanet species, clearly Mars offers us the best shot. The topography of Mars resembles the southwestern deserts of the United States, except it is devoid of visible plant and animal life. In color the surface is rusty orange and brown. Boulder-strewn flatlands are common, but their monotony is broken at frequent intervals by spectacular mountains and chasms whose dimensions dwarf terrestrial features. For example, Olympus Mons is the tallest volcano in the Solar System, three times as high as Mount Everest, and covers an area about the size of Montana. Valles Marineris, a system of canyons slicing into the planet's midriff, is more than ten times the length of the Grand Canyon and in places is twenty-four times as wide and three times as deep. As on our moon, meteor craters and volcanoes are the predominant surface features, but unlike the Moon, Mars shows clear evidence that copious amounts of water once flowed over its surface.

Scientists believe that at one time Mars had a warmer climate and a much denser atmosphere laden with water vapor. On the surface, running water carved out networks of channels, and the planet could easily have harbored life. So far terrestrial experiments have failed to create life, but biologists have been successful in creating some of life's fundamental building blocks, amino acids. They do it in a laboratory by passing an electric current (simulating lightning) through a primordial soup composed of compounds of carbon, nitrogen, oxygen, and hydrogen. Since all these elements are present in the Martian atmosphere, scientists speculate that amino acids were created during Mars's early history and that simple life forms may have appeared.

At any rate, the thick atmosphere has largely disappeared and ice-age conditions prevail today. The only water is frozen in the polar ice caps and perhaps in underground permafrost. Ice caps cover both poles and change size with the seasons. The ice, formed from both water and carbon dioxide, has been called "frozen club soda." Most of the water is at the North Pole, most of the carbon dioxide at the South Pole. Why this asymmetry, no one knows. A greater mystery is what happened to all the water and ice that gouged out those huge channels billions of years ago. Mars has a strong enough gravitational field to hold water vapor in its atmosphere rather than allow it to escape into space. Perhaps the water went underground and exists today as subsurface permafrost or even in great frozen reservoirs. Before people venture to Mars, we need to know more about where and how they might obtain water—one of our most fundamental needs.

Most of what we know about Mars comes from data collected over several decades by unmanned probes. Four Mariner spacecraft orbited or flew past Mars between 1965 and 1971, and two Vikings landed on it in 1976. The Mariner missions mapped the entire planet and revealed the heavily cratered surface that is reminiscent of our moon, yet much more varied, with splendid valleys and magnificent peaks. The Viking probes took more than fifty thousand pictures and analyzed a small amount of Martian soil. Each lander extended an arm that scooped up several tablespoons of soil and inserted it into a test chamber, where chemical and biological checks were made. Silicon dioxide composed nearly half the soil. Iron oxide made up almost a fifth, and the remainder was divided among oxides of aluminum, magnesium, calcium, sulfur, and titanium. No organic compounds were found. No carbon-based units were detected—

a great disappointment because life on Earth is based on organisms containing carbon. Plenty of carbon dioxide in the atmosphere, but no carbon detected in the soil.

The Viking biological experiments produced strange results, at least to this layman. For example, when water was added to one sample, a sudden increase in oxygen was measured. Did this mean that the moisture had activated something alive that, like a terrestrial plant, was giving off oxygen? Scientists decided not; they based their opinion on the results of another experiment that produced an increase in oxygen from soil that had been heated enough to kill even the hardiest organisms. Another experiment added nutrients dubbed "chicken soup" to a soil sample. At first, there was a large buildup of carbon dioxide and oxygen, which fell off as more chicken soup was added. Was the initial surge caused by organisms that had metabolized the chicken soup and then decided they didn't like it, or that had been killed by it? Apparently not. "Everything we see in the . . . experiment points toward chemical oxidation," said the head of Viking's biological team.

In a third test chamber radioactive carbon 14 was added to some soil. The gas in the chamber was then monitored, and radioactivity was found. Had something metabolized the carbon? Again, the scientists concluded no, that it was a chemical—not a biological—reaction, perhaps one caused by ultraviolet radiation from the Sun that had beaten down upon the Martian surface and created some strange, unearthly oxidation reactions in the soil.

I conclude from all this that Viking proved three points. First, there is no large-scale life on the surface of Mars, or the photographs would have shown it. Second, there were no Earth-like organisms in the soil samples tested. And third, if anything was alive in the samples, it was small and its biological processes are not the same as those of terrestrial organisms. But note that the second and third conclusions apply to only two specific points on the entire surface of Mars, an area about the same size as all the land masses of Earth. Two robotic spacecraft visiting Earth, the first analyzing sand in the Gobi Desert and the second ice in Antarctica, might draw some strange conclusions. I would love to prowl on Mars and dig down a bit, where ultraviolet rays have not cooked everything, especially near the polar ice caps or in the warm hollows of dry riverbeds. Discovering something there would be spectacular, it's true—but perhaps almost as valuable would be proof that Mars *is* lifeless. As Carl Sagan put it, "But if Mars *is* lifeless, we have two planets, of virtually identical age, evolving next door to each other in the same solar system: life arises and proliferates on one, but not on the other. Why? This is the classic scientific circumstance of the experiment and the control." Maybe we truly are alone in this solar system. We should find out. Maybe we are alone in the entire universe, but I can't believe that. There are just too many stars, too many planets, and it is cosmic conceit to think that out of the trillions, ours alone bears life.

CHAPTER

III

ELLIPTICAL PATHS

As Mars and the Earth orbit the Sun, their slightly elliptical paths bring them as close to one another as 35 million miles and as far away as 220 million. This relative movement of the two planets is crucial in planning our journey from one to the other, our *transfer trajectory* in the language of the space engineers. Fortunately our knowledge of planetary movement is so exact, being governed by the laws of physics, that astronomers can predict an eclipse of the Sun within a fraction of a second. They know that the Martian year is not 687 days; it is really 686.97964 days. With this precision, we can not only trace in our minds but also store in our computers various paths between the orbits of Earth and Mars. The Earth, closer to the Sun, is moving faster than Mars and overtakes it, on the average, every 780 days. Because the two orbits are not perfect circles, this interval varies from twenty-five to twenty-seven months, and forms the basis of deciding when to depart Earth for Mars (the *launch window*) and when to return.

The most economical trajectory, in terms of rocket fuel required, is an elliptical path tangent to Earth's orbit at departure and to Mars's orbit at arrival. Such a trajectory—*minimum energy* in the jargon of the astronaut or astrophysicist—is called a Hohmann, after Walter Hohmann, a German civil engineer who wrote about optimum transfer trajectories as early as 1925. Half a Hohmann ellipse, or a one-way trip, takes between six and nine months. However, after arriving at Mars on a Hohmann trajectory, our crew would discover that a similar return was then impossible, because the two planets would be in the wrong alignment. The astronauts would have to wait in the vicinity of Mars for about a year and a half until the two planets reached the proper positions for a Hohmann transfer back home. Thus a minimum-energy path both ways would result in a total trip time of more than two and a half years.

There are various ways to speed up this process, but any deviation from a Hohmann transfer results in higher energy requirements—that is, you have to get going faster, and that means more fuel. More fuel, more weight; more weight, more trouble and expense in mounting the expedition. There are, however, many clever ideas being discussed today for reducing the time, weight, and cost of a trip to Mars. One involves splitting the mission by sending a slow (Hohmann) cargo vehicle first and then following with a faster passenger craft. The two would rendezvous in Mars orbit. NASA calls this a "sprint" mission. The crew would not depart Earth until they were sure that their precious cargo (including fuel needed for the return trip) had arrived at Mars safely. Trip time for the crew could be reduced to fifteen months, including a month on the surface of Mars. Nonetheless it is disquieting to some—me included—to think that it all hinges on the two halves, separated at times by millions of miles, coming together and docking successfully. If they miss somehow, the sprint mission crew is dead.

Another clever idea involves using the gravitational field of Venus as a slingshot to speed a spacecraft on its way to Mars. Remember that Venus is closer to the Sun than Earth, while Mars is farther away. Aiming for Venus while intending to end up at Mars may seem like going off in exactly the wrong direction, but orbital mechanics, involving intricate curved paths of varying speeds, is a deceptive science. It's sort of like ski jumping: the best way to soar high is to get up a good head of steam going downhill. In similar fashion, a spacecraft heading "down" toward Venus picks up speed and retains it when it changes direction and soars back "up" to Mars. It would take about eleven months to reach Mars by way of Venus. This is slower than a Hohmann, but the advantage is that planetary alignment would be far superior and would permit a trip back to Earth after two months on the Martian surface.

Another proposal would establish some kind of "ferry" vehicle in a continual orbit between Earth and Mars. Crews and supply ships could meet the ferry near Earth and hitch a ride to the vicinity of Mars, where they would separate from the ferry, slow down, and land. Then vice versa for the return trip.

The use of staging points has also been suggested. The most obvious one is the Moon. In this scenario, a lunar base is the launch site for a Martian expedition, perhaps with lunar materials being used for part of the construction and propulsion of the Mars craft. Other staging areas could be the *Lagrangian libration points.* Named after the eighteenth-century French astronomer Joseph Louis Lagrange, these are five points in space at which the gravitational pull of Sun, Earth, and Moon are exactly balanced. Two of them, L-4 and L-5, are stable, which means that an object placed there will stay there and will tend to return if nudged away. L-4 and L-5 are located along the Moon's orbital path, one 60 degrees ahead of the Moon and the other 60 degrees behind it, and each point is equidistant from Earth and Moon. The idea of these points has intrigued people for a long time, and they have become symbols of space exploration. For example, during the 1980s a ten-thousand-member L-5 Society thrived in this country. The use of staging points does not reduce the total energy requirements of a trip to Mars, but it does separate the mission into more manageable phases, each of which can be serviced by smaller, more specialized machines. Whether the whole is greater or less than the sum of its parts remains to be seen. There are hundreds of intricate, interlocking factors that have to be analyzed before one can chart an exact course to Mars.

In an attempt to bring some order out of this chaos, in this book I will discuss one specific trajectory. It may not be the one Martian travelers will actually use, but it does represent reasonable compromises between competing approaches and is consistent with today's technology and the advances we might expect by the turn of the century. Our mission departs Earth on June 3, 2004, takes a little more than five months to swing by Venus, and arrives at Mars on

May 9, 2005. This leg of the trip takes longer than a Hohmann transfer, but planetary position is better upon arrival, and departure can then be scheduled for July 8, 2005, avoiding a year-and-a-half wait. During the two months in the vicinity of Mars, probably forty days will be spent on the surface, with the remainder in orbit preparing for the landing and for the homeward journey. Thus our total trip time will be twenty-two months. The launch date was selected based on optimistic assumptions: I doubt that a Martian expedition will actually depart Earth as early as 2004, but it is within our technical capability to do so. This book discusses some of the factors that determine when it will happen. That it will be done, sometime by someone, is beyond doubt.

CHAPTER

IV

PRECIOUS MATERIALS

THERE IS no doubt that a Martian expedition is an undertaking of unprecedented complexity. In Earth orbit an astronaut in trouble can press his retrofire button and be back on the surface within an hour. Even the Moon is no more than three days away. But once committed to a Mars trajectory, it will be very difficult for the crew to turn back in the event of an emergency. After several months, the outward-bound craft will be too far away to reverse course, and will have to continue past Venus before returning to Earth. A voyage of twenty-two months will certainly impose new standards of performance on both crew and machinery. Reliability will become all-important, and both crew and equipment will be tested exhaustively. No hospital, no repair shop, just month after month of black sky and, at times, a friendly face on the TV screen.

A little arithmetic demonstrates the enormity of the problem: the average man consumes daily about 3 pounds of food, 7 pounds of water, and 2 pounds of oxygen, for a total of 12 pounds. Most women consume less, a point in their favor in planning any long-duration mission. To support a crew of eight on a twenty-two month trip, 65,000 pounds of food, water, and oxygen would be required. More than half of this weight is water, and that includes only water for drinking, with no allowance for showers or laundry—clearly an intolerable situation. On Earth, waste water is processed (by sewage treatment, filtering through soil, evaporation, etc.) and used again. On a Mars expedition, similar recycling will be required; if all wastes were simply dumped overboard, the weight of all the fresh water would be an insurmountable problem.

Of course, the weight of the crew's consumables is just the tip of the iceberg. When Neil, Buzz, and I left Earth orbit and headed for the Moon, our weight was roughly 100,000 pounds: 65,000 for the round-trip mother ship and 35,000 for the lander. Most of this weight was fuel. To accelerate all this weight from launch pad to escape velocity required a huge Saturn V booster weighing 6 million pounds. Big numbers, just to land two men on the Moon and return three to Earth. For Mars, the departure velocities from both Earth and Mars are higher, and all the machines will be a lot bigger, so that about two dozen boosters the size of the Saturn V will be required. The parts of the Mars craft will be assembled in Earth orbit and then sent on their way. I calculate that outward bound, the total spacecraft weight will be at least a million pounds. That's ten times the weight of the Apollo craft, not surprising considering the larger crew, the more complex tasks, and the immense distances involved in a Venus-Mars mission.

The cost of a Mars expedition is more difficult to estimate than the weight, although the two are not unrelated. Today the major citizens' group pushing for a Mars mission is the 125,000-member Planetary Society. The Planetary Society has sponsored a private-industry report that concluded the cost of a piloted mission would be $40 billion, in 1984 dollars. The Apollo Moon program cost, during its time, about $24 billion—or $75 billion inflated to 1984 dollars. In other words, one Mars landing would cost considerably less than Project Apollo.

Part of the problem in making comparisons such as this is knowing what to include—and what to leave out. For example, must we include the cost of developing a heavy-lift rocket, or would one be available because of military requirements or other NASA programs? Where would the Mars machinery be assembled—at an Earth-orbiting space station? If so, must we count the cost of designing and launching that station, or would we have it anyway, even if we didn't want to go to Mars?

There are other estimates to compare with that of the Planetary Society. The report of the National Commission on Space, issued in 1986, describes a comprehensive exploration of the Solar System, including a Mars landing by 2015 and the establishment of a colony there by 2028. The price tag to sustain this program (Mars plus a lot of other things) is estimated at around $30 billion per year. A NASA study chaired by former astronaut Sally Ride concluded that the landing flight itself would be $45 billion, and that supporting expenditures would drive the total bill to $95 billion. A Soviet expert, Roald Sagdeev, thinks that a major manned expedition would cost $50–100 billion, "assuming the dollar will be stabilized." This puts Mars within the range of the B-2 bomber program ($70 billion) and still considerably less than the bailout of the savings-and-loan industry.

In 1986 a group of thirty-four students at the University of Michigan devoted a semester to planning Project Kepler, a theoretical manned Mars mission. Their estimated cost: $22 billion in 1984 dollars. The students tried to express this abstract number in ways that would bring it a little closer to home. First they spread the cost over fifteen years. Then they divided it by the population of the United States, 235 million people. Their conclusion: it would cost each American $6.33 a year for fifteen years to develop and carry out a Mars mission. The students compared this $6.33 with the per-capita expenditure for cigarettes, $93.01, and potato chips, $12.46.

My own conclusion is that, considering the technological unknowns, all these estimates are probably too optimistic, even Sagdeev's top figure of $100 billion. Perhaps that number should be doubled. The $200 billion could be spread over fifteen years, as the Michigan students point out, although to make the 2004 launch date we have to start spending now. Our potato-chip budget would definitely be exceeded—all the way up to $57.50 per capita per year—but we'd still be spending more on cigarettes. Such numbers, however, make sense only in the context of one Mars mission—one landing and one return. If the first flight is considered just a trailblazer, leading to the eventual colonization of Mars, how meaningful is its price tag? Was

Columbus's purchase of the *Niña, Pinta,* and *Santa María* cost-effective? The only thing I know for certain is that starting a human colony on a second planet will cost much less than the weapons we buy to destroy the first one.

CHAPTER

V

HAZARDS OF THE DEEP

THE SPACE environment between Earth and Mars is fundamentally the same as it was during the Apollo forays to the Moon. There is just more of it. The hazards are the vacuum of space, unforgiving of leaks; meteorites that can create those leaks; and the constant bombardment of particles coming from the Sun and of gamma rays from intergalactic space.

First, there are a lot of dust particles in space. On Earth we see them as *meteors*, when they enter the atmosphere and burn up—brief streaks of light across the night sky. The Perseid meteor showers in August and the Geminids in December are highly visible visitors. Before a meteor enters the atmosphere it is called a *meteorite*; if it survives the atmosphere and strikes the ground it becomes a *meteoroid*. A very few meteoroids have been found that weigh thousands of pounds. If one of these had hit a spacecraft before plunging to Earth it would have been like the iceberg that sank the *Titanic*, only a lot quicker and a lot more violent. In an ultra-high-speed collision the meteorite does not have to be very large—one with the diameter of a dime would do it. An explosion, then depressurization and death. Fortunately, in all the years of manned spaceflight, no craft has been hit by a meteorite large enough to cause damage. On the other hand, virtually all spacecraft do encounter dust particles, or *micrometeorites* as they are usually called, and their impact causes tiny craters on the skin or exposed equipment of the craft. Usually no damage is done, but optical surfaces such as telescope lenses are an exception. They gradually become pitted and harder to see through.

The likelihood of getting zapped by a large meteorite of course increases with exposure time. Planners of a long voyage to Mars and back must certainly consider the possibility, but not overreact, because no matter how thick the spacecraft walls, one can always hypothesize a meteorite large enough to penetrate them. Astronomers believe that most meteorites are fragments of disintegrated comets, in orbit around the Sun. Others may come from the asteroid belt, located between Mars and Jupiter, which means there may be more meteorites near Mars than close to Earth because of Mars's proximity to the asteroid belt. The U.S. space station *Skylab*, which orbited the Earth between 1973 and 1979, was designed to include a meteorite shield, a thin metallic sheath that at launch was flush against *Skylab*'s skin but that once in orbit popped outward a few inches. The idea was that a meteorite striking the shield would break into harmless fragments before reaching the spacecraft itself. The shield also doubled as a parasol, keeping the interior of *Skylab* comfortable despite a peak temperature of 250 degrees Fahrenheit at the outer surface of the shield. Our Mars craft should probably have similar protection.

A more serious design problem involves protecting the crew from radiation. We Earthlings are shielded by our planet's magnetic field and atmosphere from most solar and intergalactic radiation. As our Mars crew ascends to Earth orbit they will be below two bands of radiation known as the Van Allen belts (after University of Iowa physicist James Van Allen). Trapped by Earth's magnetic field, the inner belt is composed of protons, the outer of electrons. Leaving Earth orbit, our crew will pass through the Van Allen belts quickly, as we did on our way to the Moon, so that only a small dose of radiation will be absorbed. But in deep space—between Earth and Mars—our spacecraft and crew will be constantly exposed to several different types of radiation, one of them potentially lethal.

Actually, *ionizing* radiation is a more precise term because radio waves and light are also forms of radiation but are harmless in that they cause no ionizing reaction within the human cell. Cells react to harmful radiation by producing toxins and by abnormal cell division or death of the cell. In acute cases, human response can range from nausea and vomiting to fever and death. Long-term effects, which may not occur until years after exposure, include cataracts, tumors, and leukemia. Pregnant women have to be especially careful because a fetus bombarded with ionizing radiation may develop into a baby with monstrous defects—a process called *teratogenesis.*

Intergalactic cosmic rays are very high-energy particles

(protons and heavy ions, such as iron) that originate somewhere outside the Solar System, probably in the explosion of stars called supernovae. Although relatively small in number, their presence is constant. They can easily penetrate the walls of a spacecraft and pass through the bodies of the crew. When they collide with something solid, like the aluminum walls of a spacecraft, they can produce nuclear fragments, such as neutrons, that cause dangerous cell ionization and thus add to the dose the crew receives.

Then there are solar flares. This radiation, consisting primarily of protons, belches out sporadically from the Sun. The frequency and intensity of storms on the Sun vary in an eleven year cycle, but even during a quiet year the Sun can release unexpected bursts of energy that would kill a Mars crew—if unprotected—within days. Although solar flares usually last only a day or two, a single event can hit a crew with more radiation in a couple of hours than cosmic rays could deliver in a decade. The last period of maximum solar activity was in late 1979. The next is expected in 1990, and our Mars mission of 2004–06 can expect to be some what past the peak of the eleven-year cycle.

A Mars ship also has the capability of producing its own ionizing radiation, if it carries a nuclear power source for propulsion or for generating electrical power. Nuclear reactors produce neutrons and gamma rays, and the crew must be shielded from them as well as from radiation coming in from outside.

Radiation is generally measured in *rems*, a unit that takes into account not only the amount of ionizing radiation but also its biological effect. Analyzing risk is a difficult task. The young, and especially pregnant women, need more protection than those whose reproductive years are over. The U.S. government allows workers exposed to radiation to absorb 5 rem per year, and NASA has set a career total of 200 rem for astronauts. In 1979 a solar flare was recorded that would have caused an exposed crew to receive 2,600 rem—about seven times a lethal dose. Clearly a great deal of shielding will be needed to protect our Mars crew.

Fortunately aluminum, the most practical material for a spacecraft pressure hull, makes an effective shield. A couple of inches of aluminum is sufficient to protect against intergalactic gamma rays and all but the most severe solar flares. For those, extra protection is required. Our Mars mission will carry large quantities of water and fuel, and both are good energy absorbers. When notified of a solar flare, the crew will maneuver their craft so that the fuel and water tanks are between them and the Sun. Furthermore the crew must have its own early-warning system on board, because at certain times during the flight the position of the Sun will cause communications with Earth to be interrupted.

On the surface of Mars humans will not be protected as they are here on Earth: Mars lacks a magnetic field, and its atmosphere is too thin to be an effective shield. Not counting solar flares, a person on Mars would receive about 1 rem per month—not bad. But again, protection from solar flares must be provided, either by taking shelter in the spacecraft or by creating barriers out of Martian soil. In addition, a nuclear power source will have to be designed to operate at some distance from the crew—and preferably behind a boulder.

Although predictions are difficult until an actual spacecraft has been designed, built, and tested, experts feel that on our first twenty-two-month round-trip, the crew will absorb no more than 100 rem. This amount would make them a couple of percentage points more likely to develop cancer later in their lives. Of course it is also theoretically possible that one giant solar flare could wipe them out, especially if it caught them unawares.

22 Month Rounder

CHAPTER

VI

WEIGHTLESS

OUR CREW must be protected not only from meteorites and radiation but also from weightlessness, which when experienced over a long period of time can have debilitating effects on the body. On Earth our bodies must constantly fight against gravity: muscles strain as we walk up stairs, and bones maintain their strength to support the weight of our bodies. On a trip to Mars, floating free, the crew's bodies will gradually change in an environment unlike anything encountered by their ancestors during millions of years of evolution.

The first change will be in the distribution of body fluids, which on Earth are drawn by gravity toward the feet. Muscle contractions around the veins of the legs counteract this tendency, and so do one-way valves in the veins, by closing between heartbeats. In weightlessness, fluids migrate toward the upper body. The astronaut notices this right away, sensing a fullness in the head, a stuffiness in the nose, and perhaps reddened eyes. During my flight to the Moon, Armstrong and Aldrin looked different to me, as if they had received an instant face-lift. Because of the lack of gravity and the resulting fluid shifts, wrinkles disappeared and their eyes looked squinty and crafty. The three of us also grew by an inch or two with no gravity to compress the space between our vertebrae. These changes are trivial and will not harm the performance of our Mars crew.

Although it didn't happen to Neil, Buzz, or me, weightlessness makes about half of all space travelers feel sick for a couple of days. The symptoms are similar to seasickness, with loss of appetite, sweating, nausea, and sometimes vomiting. Exactly why this happens no one knows, but one theory is that the brain receives conflicting messages from the eyes and ears. In the inner ear, weightless fluid may cause the tiny hairs, or cilia, to wave back and forth, sending a message to the brain that says, "Things are in some kind of strange motion here." The eyes, on the other hand, report no such movement. Why the brain reacts to these contradictory signals by emptying the stomach, I have no idea.

Other changes are equally mysterious. For example, the body mistakenly senses increased fluid in the thoracic region as an increase in the body's total blood volume and initiates a complex process to reduce it. Thirst is reduced, and hormones are secreted that have a diuretic effect, causing increased urine production and a decrease in blood plasma. The body then considers that this new volume of blood is too rich in red cells and sets about curtailing the production of new ones in the bone marrow. Of those cells that *are* produced in weightlessness, some are different. Under a microscope, Earth-produced cells are normally disk shaped, like doughnuts but with a thin center instead of a hole. But their space cousins look ragged around the edges, somewhat resembling a starfish. These irregularities disappear after return to Earth; in Martian gravity, who knows? However, this does not seem to be a serious problem.

Muscles atrophy if not used; without exercise leg muscles would wither during a long space voyage, such as a trip to Mars. Nor is the heart muscle immune. With no gravity to pump against, the weightless heart grows lazy and shrinks in size. This loss of muscle mass and tone is not a disease but a natural result of the body's adaptation to a new environment. In all likelihood there are no harmful consequences—as long as the body remains weightless. But upon reaching Mars or returning to Earth, muscles may not be up to the simplest tasks, such as walking or just standing. Space travelers have suffered from a condition called *orthostatic hypotension*—low blood pressure due to body position. When they return to Earth and first stand up, gravity forces blood from their heads and upper bodies down into their legs. Flapper valves in the central leg veins normally close between heartbeats to prevent blood from falling back down. This is unnecessary in weightlessness, and it takes the autonomic nervous system a little while to relearn this trick after returning to Earth. In addition, in space the leg muscles, which should tighten and constrict these veins, relax too much and allow the veins to distend. Consequently after long-duration flights some cosmonauts have felt lightheaded and have been unable to walk properly for a day or so. Their leg strength and their sense of balance may also take some time to return to preflight norms. Landing on

Mars—a planet with only 38 percent of Earth's gravity—will definitely be less taxing physically, but the crew may not be able to perform at their peak for several days.

So far the most serious physical problem known to be caused by weightlessness is bone demineralization. Just as the marathon runner does not need the heavy bones of the weight lifter, so the space traveler's skeleton can be much lighter than that of his Earth-bound twin—especially the bones of the lower back, legs, and feet. Exactly how the body senses this remains a fascinating mystery, but somehow it knows and responds to weightlessness by excreting calcium, thereby reducing bone density. So far the loss has averaged nearly 0.5 percent per month. A high-calcium diet may help a little, but not much. As with elderly people who develop osteoporosis, the bones of a space traveler may become not only light but weak and brittle. Bones on Earth begin to fracture spontaneously when their density is reduced by 30 percent, a condition that would take about five years to develop in weightlessness, according to NASA's data. It seems to me that on the surface of Mars, working hard inside a pressure suit, an astronaut might be susceptible to fractures at considerably less than this 30 percent level. Tests of bed-rest patients—a poor simulation of long-duration weightlessness but the best we have—suggest that after one year some bone damage may be irreversible. The skeleton contains two kinds of bones: light, honeycombed trabecular bone and denser cortical bone. It is the loss of trabecular bone that might be permanent.

As is so often the case with the human body, one abnormality can be accompanied by another, and yet another, in a domino effect, or at least an interlocking matrix of effects. For example, loss of calcium is frequently accompanied by a loss of magnesium, which in turn can alter the chemical balance of heart muscle and result in irregular heartbeats. The possible formation of kidney stones is also more apt to be a problem, as the lost calcium is excreted through the urine. Anyone who has ever experienced one of these jagged little rocks (as unfortunately I have) knows that it can cause excruciating, incapacitating pain as it wedges its way through a tiny, tender tube, headed for freedom—you hope—in the bladder. If it gets totally stuck in that tube, called the ureter, urine can back up in the kidneys and cause a life-threatening situation. On Earth kidney stones can be disintegrated by vibration or removed by surgery, but it takes a skilled physician and fancy equipment that might not be practical to carry to Mars. Certainly someone like myself, a proven kidney stone producer, should be barred from a Mars crew.

Of course the preflight physical screening will consider not only kidney function, but literally thousands of other factors. Appendicitis is another real possibility, and preventing it by removing the appendix is certainly worth considering. Once I asked Dr. Joe Kerwin, a *Skylab* astronaut and physician, what he would have done if one of his mates had shown symptoms of appendicitis. "Strap an ice bag on his belly, give him some antibiotics, and come on home," he replied. On Mars the ice bag and antibiotics might have to suffice, for better or worse. I've always considered preventive surgery silly for a short trip to the Moon, but not for a Mars mission.

Full body scans—CAT scans, X ray, etc.—will also be valuable before the flight in detecting small tumors that might grow to fatal proportions in twenty-two months. No doubt about it, our Mars crew will be poked, pummeled, probed, and pierced like no other, but as Dr. Arnauld Nicogossian, NASA's director of life sciences, has pointed out, "Never before has medicine been called upon to certify that an individual will be healthy enough to perform for two years following the examination."

Although the Mars spacecraft will be assembled in a "clean room," where dust particles have been filtered out, nonetheless microbes pervade all equipment and undoubtedly a colony of them will accompany our crew on their voyage. In all likelihood no virulent strains will find their way on board, but it is even possible that in the alien environment of space, with weightlessness plus plenty of radiation, some genetic mutations might develop and produce new forms of bacteria that humans have never encountered before. Just as Columbus had to fight scurvy and syphilis, so may the first Mars crews find that disease awaits them far from their home port.

Of course scurvy is easily cured by vitamin C, and our knowledge of nutrition today permits explorers to enjoy a balanced diet on mountaintops or deep under the sea. The food we carried to the Moon on the Apollo flights was certainly nutritious, but it was not very tasty. For a week or so, people don't care, but for a Mars expedition food will transcend matters of nutrition and weight control. Food—three times a day over the lonely, monotonous miles—may be the key factor in preserving the morale of the crew, in keeping them functioning at a high level of efficiency. The wardroom table will be not only a place for apportioning calories but the social center, with meals as the catalyst that brings conversation, rapport, and the friendly discussion of problems that might otherwise be bottled up, later to explode.

Good food can be a wonderful elixir, no doubt about it, but as with Jack Sprat and his wife, people can disagree totally on its selection and preparation. "What is food to one, is to others bitter poison," according to Lucretius, the Roman poet and epicure. Some people are finicky eaters, and some change their mind periodically about various foods. Even the color of food can be important, especially the *lack* of color in most rehydrated foods. The three crews who lived aboard *Skylab*, the world's first space station, found that dishes they remembered as tasty on the ground were bland in space, and they suggested that an array of hot sauces be carried on future flights. Although individual desires were considered on *Skylab*, the menu tended to be monotonous. Astronaut Jack Lousma lamented during his flight, "I've asked myself every six days, whenever it turns up on the menu, 'How come I picked beef hash for breakfast?' " On our Mars trip, Jack would be exposed to beef hash for breakfast more than a hundred times. If he started hating it after only a month, after a year he might end up throwing it around the cabin. Jack's crewmate Owen Garriott, normally not a complainer, said that the beef dishes tasted as if insecticide had been added to them. Everyone on *Skylab* missed fresh food.

The basics of a healthy space diet are well known: slightly less than three thousand calories per day, with the emphasis on complex carbohydrates and protein. Fats should be carefully controlled, with mono- or polyunsaturated fats preferred to saturated ones, such as animal fat or palm oil. The link between high serum cholesterol and heart disease is too well documented to be ignored when twenty-two months away from a cardiac intensive care unit. The list of vitamins and minerals is well known also. Vitamins A, B_1, B_2, D, thiamine, riboflavin, niacin, and ascorbic acid will be provided. Likewise minerals: calcium, iron, phosphorus, iodine, copper, magnesium, zinc, manganese, molybdenum, and potassium. Especially calcium, because of its loss from the skeleton during weightlessness.

In the early days of the space program, we had two kinds of food: bite-sized cubes (compressed bacon, peanut-butter cookies, etc.) and tubes of dehydrated food and drink to which we added water. It was only cold water at first; then on Apollo we had our choice of hot or cold, a big improvement. Cold coffee has never appealed to me. By the time of *Skylab*, in 1973, things were a lot more civilized, and a freezer was added for such items as steak and ice cream. The shuttle features not only hot and cold water dispensers, but a pantry, an oven, food-serving trays, and a place to clean up the mess. The shuttle has a menu of seventy different food items and twenty beverages, all of them nonalcoholic. The food is preserved in various ways, and may be freeze-dried, irradiated, dehydrated, or thermostabilized. Some foods, like nuts, need no processing, but most must be protected from decay. For example, on the shuttle beef and turkey are packaged in foil pouches that have been exposed to ionizing radiation. Standard eating utensils are used, and the food can be seasoned with serving-sized packets of hot sauce, catsup, mustard, and even salt and pepper immersed in a liquid. This last is a precaution to prevent particles from floating around the cabin, getting into eyes, lungs, and equipment. Also the food must be mushy enough to stick to the plate and not wander off.

On Mars there will be a highly desirable extra—fresh food. In a Martian colony it will be feasible to grow plants and animals (strawberries for breakfast, rabbit for dinner), but in the tight confines of our spacecraft nothing on that scale will be undertaken. Both the United States and the Soviet Union have grown plants under weightless conditions, with some strange results. For instance, mung beans have germinated successfully but then become disoriented and, lacking an "up" or "down," have sent their roots out helter-skelter, some of them above ground. Scientists don't yet know what makes a plant's roots grow down and its stem up, a response they call *gravitropism.* A later chapter will discuss the difficulties of providing for the crew's needs, but suffice it to say here they will not have to depend on growing food to survive. Some plants, and even a few flowers, will be grown, but more for morale than nutrition. The crew will be adequately fed, but still . . . cosmonauts have been reported to have developed sudden cravings for honey and apricots. About all Mission Control will be able to tell our Mars travelers is, "You should have thought of that two years ago."

More than fifteen thousand meals will be served on our Mars expedition, a gigantic undertaking in itself. The inventory alone—keeping track of what's where, what's already been eaten, and what remains—will keep the cook busy. Computers will no doubt assist. How about sirloin steak tonight? "No," answers the computer, "quantity zero." What's available? "The corned beef hash is highly recommended, monsieur."

CHAPTER

VII

MEDICINE AND THE MIND

COSMONAUT VALERI RYUMIN, contemplating life aboard the *Salyut* space station, wrote in his diary, "If you want to encourage the craft of murder, all you have to do is lock up two men for two months in an eighteen-by-twenty-foot room." Prisoners at least get out into the exercise yard, but for twenty-two months our Mars crew, eight strong—or weak—will have to endure one another's company whether they like it or not.

In the early days of spaceflight, the medical community speculated widely about the effects this new medium might have on a person's psyche. Dr. Charles Berry, who later was billed as the astronauts' personal physician, summed up one school of thought: "The psychological problems presented by the exposure of man to an isolated, uncomfortable void seem to be more formidable than the physiological."

The physiological could have been bad enough. Some doctors hypothesized that people in weightlessness might suffer from a long list of bizarre calamities: they would not be able to swallow, their bowels wouldn't work without gravity, their hearts might cavitate like a pump, beating wildly and churning up a bloody froth.

Beyond these physical horrors lurked the dark domain of the mind. It was possible that, in some way unknown to medical science, the operation of the brain might be gravity dependent. In weightlessness, the brain might simply shut down, and its owner would cease to function. Short of this there were other possibilities. One popular comparison was with ocean divers, who occasionally developed a "rapture of the deep." So overcome by their glorious surroundings (or so poisoned by the mixture of gases they were breathing), these divers would lose all judgment and attempt to go deeper and deeper, until eventually they killed themselves. Might not space travelers, this new breed called astronauts, be overcome by their sudden view of infinity and want to penetrate it, abandoning their emotional attachment to Mother Earth?

The early astronauts quickly punctured all these balloons. Their bowels and brains functioned, and they were incensed at the notion that their minds and bodies might fail them, as they had never done through thousands of hard hours of test piloting. They were a tough, pragmatic bunch. Their attitude was "Rapture of the deep, my ass! Let's get the medics off our backs so we can get on with our business, testing these new machines. Sure, they fly above the atmosphere instead of through it, but so what?"

It was at this point that I signed up, becoming one of a group of thirty astronauts in 1963. To varying degrees, we all regarded the medical community with suspicion. The flight surgeon could tell us to take our clothes off and get up on that table, where he might find things wrong that would ground us—in an instant destroying a lifetime of dreams. The medics were the enemy. We particularly objected when they intruded into matters we deemed "operational," a term ill-defined but centering on how flights would be conducted and how decisions made, and by whom.

As time went by, with more and more successful spaceflights, the tensions subsided somewhat. Astronauts are not stupid, and they could read the evidence of bone demineralization as well as anyone. Then astronauts were selected who were also medical doctors, and that helped a lot to bring the two viewpoints (test pilot versus medical guinea pig) closer together. By the time of *Skylab*, with three men staying up for eighty-four days, at least an uneasy truce existed, if not a spirit of all-out cooperation.

But that was in regard to the physical procedures: blood and urine tests, exercise and dietary protocols. I have always regarded the mental dimension of spaceflight as something that will work itself out, without any help from psychiatrists or psychologists. Crew selection and training have always been "operational" matters. On *Skylab* hostilities did develop at times, but they were between crew and ground personnel, not among crew members. Assembling and training a crew is a long and arduous process that creates plenty of tension. The crew commander can spot trouble and is in the best position to decide what to do about it. At least that is what I have always felt. And the record backs up that point of view: after more than fifty flights, the American space program has experienced no psychological problems in flight.

But Mars is different. The flight time is the main factor, but to it must be added the distance, the unprecedented cosmic separation from the home planet. It won't be the same as the long-duration flights in Earth orbit, where the crew members feel a kinship with Earth because of the bright, ever-changing panorama of the planet out their window. It won't be the same as a lunar voyage, where home is still large enough and beautiful enough to act as a strong magnet for the eye and the mind. On a Mars flight, when Earth fades to nothing more than a very bright star, things will be different.

The psychological burden will come in two parts, the double whammy of isolation and confinement. Isolation from everything the crew member has known before: no family, no trees, no valleys, no waterfalls. Out the windows, month after month, the same black velvet dotted with unblinking stars. As the mission unfolds there will be only slight changes in the Earth, Moon, Sun, and Mars—and these will not be noticeable on a daily or even weekly basis. On top of that comes confinement, with no way to escape from the cloistered daily life of the spacecraft. No way to avoid looking at the same walls, no way to escape the same repetitive tasks, no way to get away from other crew members.

The interaction between individuals may be the most crucial factor of all, and the psychological climate can change drastically as months of boredom create frustration and hostility, perhaps followed by incapacitation, violence, or even insanity. For the first time in spaceflight, "group dynamics" may no longer be psychobabble but a life-or-death matter, and the key to a successful mission. If I were going on the trip, I would not be as suspicious, even disdainful, of psychiatric assistance as I once was. I'm not quite prepared to wave a white flag of surrender at the medics, but I am willing to admit that their time has arrived, finally, that their concerns and precautions may be valid for the first time, and a Mars crew ignores them at its peril. Take my appendix out; give me crew members whose company I will enjoy, or at least tolerate. Mars is too far away to apply the old yardsticks.

Later in this book I will be very specific about what kind of crew I think will make the first journey to Mars, but for now suffice it to say that in all likelihood the mission will be an international one, and that further complicates the situation. Differences of language, culture, traditions, and beliefs will add an extra dimension to crew organization, training, and interaction. Both men and women will be included, solving some problems and probably creating others.

I think the fundamental attributes of successful crew members will remain constant. I remember being on an astronaut selection board in 1967, picking people for Project Apollo and for *Skylab*. There were four basics we looked for: intelligence, education, work history, and motivation. The first three were relatively easy to document and evaluate, but motivation was a more elusive quality. Almost anyone can rise to the occasion, in this case a forty-five-minute interview, but beyond that, who will succeed? Years of classroom instruction and individual study, thousands of hours logged in simulators, brutal work schedules, little time off: all of that faces the astronaut *before* the flight, in this case a flight of extended peril and total isolation in cramped quarters. Who will flower, who wither? Whose exterior is shiny and strong but—like high-carbon steel—shatters easily?

Consider just one factor—sex. NASA psychologist B. J. Bluth likens a short spaceflight to a date, a long one to marriage. Maybe a Mars crew should consist entirely of married couples. An element of stability, of old-shoe comfort, would be introduced by having one's husband or wife to fall back on. Certainly a singles-bar atmosphere, a charged mixture of sexually unattached competitors, would be a disaster. An all-male or all-female crew would alleviate some of these problems but would clearly introduce others. In the final analysis, sex may be ignored and the selection made on the basis of competency alone. I don't know the answer to these questions, but they are real. To the basic four individual qualities (intelligence, education, work history, motivation) I would have to add compatibility as a prime requisite in assembling a Mars crew—and I would look to the medical community for help in making these judgments.

CHAPTER

VIII

CLOSING THE LOOP

THE CREW for the first Mars flight should be as small in number as possible because of the tremendous support system each individual requires. One crew member will consume about four tons of food, water, and oxygen during the trip. This weight includes only those substances ingested into the body and is just the tip of the human-support iceberg. The *Skylab* space station, for example, carried six tons of water to sustain two man-years of occupancy. At that rate of consumption, *each* Mars crew member would require nearly six tons of water. Water weighs eight pounds per gallon, and even a quick shower can easily use a gallon, not to mention the water needed for laundry and general cleanup.

Crew members have a long list of other requirements.

Most fundamental is that because blood boils in a vacuum, humans must live inside a sturdy, reliable pressure shell. Here on Earth we are accustomed to an atmosphere of 80 percent nitrogen and 20 percent oxygen at a sea-level pressure of 14.7 psi (pounds per square inch). In the early days of the American space program (Mercury, Gemini, Apollo), weight and complexity were reduced by using a 100 percent oxygen atmosphere of only 5 psi. The shuttle shifted to a sea-level mixture of nitrogen and oxygen. That is preferable for long-duration flights and will probably be used on a trip to Mars. Inside the pressurized spacecraft the temperature and humidity should be regulated within fairly narrow limits of 65 to 72 degrees and 30 to 70 percent. Noise and vibration must be held to reasonable levels. Light must be provided. Each crew member needs space to relax and to sleep as well as adequate workstations. A variety of nutritious food must be provided. In short, humans create enormous design problems for the engineers, and their solutions inevitably involve added weight, which in turn leads to greater complexity and cost. Therefore no extra people will be carried on the first flight.

The situation is somewhat analogous to the Apollo lunar landings. Clearly one person had to stay in lunar orbit in the command module while his compatriots were on the surface. Landing on the Moon and exploring it were deemed to be too much for one person, so two were sent. For our Mars mission the tasks will be of greater complexity and much longer duration. I don't think it is a good idea to have one person alone during key mission sequences because there is too great a possibility of one person becoming ill or simply making a mistake that he or she might not catch but a companion would. For this reason four people will be required on the surface (two in the landing craft and two out on the surface conducting extended excursions), and two in orbit overhead. However, instead of choosing a crew of six I have added two others to stay in Mars orbit because I believe the overall complexity of the mission and associated equipment will require the skills of eight people to master. It's a great temptation to try it with six, but I think eight gives an extra margin, particularly because of complicated experiments in Mars orbit and because there will be two large round-trip vehicles, each containing a variety of highly technical mechanical and electrical systems. A crew of four aboard each makes sense, to me at least.

But eight people times six tons of water equals 96,000 pounds of water alone, if we do it the *Skylab* way. By that I mean that *Skylab* and all our other spacecraft to date have had open-loop systems, in which water and other wastes are stored or dumped overboard but never used again. Our Mars voyage cannot afford that luxury or inefficiency. To save weight, a closed-loop system is required, one that will recycle water, oxygen, and perhaps even some food. NASA calls such a system a *controlled ecological life-support system,* or CELSS. Much theoretical work and some experimentation have been done on CELSS designs of one kind or another, but none are near the maturity required for a Mars flight. Imagine the plumbing getting stopped up 50 million miles from home!

On Apollo spacecraft, oxygen was stored in tanks as a liquid at the very cold, or cryogenic, temperature of −297 degrees Fahrenheit. Cryogenic hydrogen, at the even more frigid temperature of −423 degrees, was also carried. The oxygen was warmed up to a gaseous state and then used to pressurize the cockpit, where the crew breathed it. In turn, the crew exhaled carbon dioxide. (As oxygen disappeared into the crew members and out through tiny leaks, it was replenished from the storage tanks.) Fans pulled the cabin gas through purifying canisters containing lithium hydroxide and charcoal. The lithium hydroxide absorbed the carbon dioxide and the charcoal cleansed the oxygen of odors. Pure oxygen was then pumped back into the cabin and the process continued. Every few hours a lithium hydroxide canister would become saturated with carbon dioxide and the crew would replace it. The old one was simply stored and returned to Earth as useless baggage. At four pounds per canister, to do the same on our Mars mission would mean storing more than ten tons of dead weight.

Oxygen was also used, along with hydrogen, to generate electricity in a clever device called a *fuel cell.* Inside the fuel cell one oxygen and two hydrogen atoms were combined to form a water molecule (H_2O). This process produced not only electrons but pure water for the crew to drink. Thus the fuel cell saved a lot of weight compared with storage batteries, not only because it was intrinsically lighter but because it eliminated the requirement to carry drinking water, except for a small emergency supply. On Apollo, beverage powders and dehydrated food were prepared by adding fuel-cell water to small plastic pouches. Waste products were stored in bags (for feces) or dumped overboard (urine). Aboard the *Skylab* space station, lithium hydroxide canisters were replaced by a fancier device for getting rid of carbon dioxide, called a *molecular sieve.* But even on a modern machine such as the space shuttle, all waste products are either stored or dumped, but never recycled.

The easiest substance to recycle is water. Excess cabin moisture caused by the crew's exhaled breath can be trapped and purified for further use. Urine, unpleasant as it sounds, can be transformed into drinking water. Carbon dioxide is a more difficult problem, but fortunately, in a common terrestrial process called photosynthesis plants use sunlight and carbon dioxide to produce oxygen, which of course the crew needs to breathe. Biological processes such as this tend to alarm engineers, who are accustomed to working with valves, gears, and other clean, tidy, well-understood mechanisms, not vats full of green crud. And of course human excrement is messier yet. In some countries it is used to fertilize crops, but not in the United States. It is filthy, teeming with bacteria, and any use of it aboard a small sealed spacecraft had better be well understood, ultrareliable, and a proven weight saver.

Fortunately, people are beginning to study these problems. In one Mars-related experiment called Biosphere II, eight people are scheduled to enter a sealed habitat near Tucson, Arizona, and remain there for two years. The crew size and duration are remarkably similar to our Mars mission. Barring emergencies, the only contact with the outside world available to the Biosphere II crew—they call themselves Biospherians—will be audio and video links, again like a Mars mission. Biosphere II is being funded by a private company, with expert help from a few government organizations such as the Smithsonian Institution. Most of the money so far has been put up by a Texan, Edward P. Bass (who has also bankrolled a replica of a fifteenth-century Chinese junk that is used to study undersea life).

Whereas most projects start small and grow, Biosphere II is taking the opposite tack. It is immense, covering 98,000 square feet of floor space and enclosing a volume of 5 million cubic feet. It is divided into sections called *biomes*, beginning with a five-story white domed building called the habitat biome. Connected to it, in turn, are an agricultural biome and then a series of wilderness biomes: tropical rainforest, savanna, fresh- and salt-water marshes, desert, and an ocean thirty-five feet deep. Of course it would be impossible to put something as huge as Biosphere II into orbit, much less send it to Mars, but the idea is first to learn how to close the various life-support loops without regard to size, and then to consider how to miniaturize them. Every part of Biosphere II must be not only in harmony with the others but in perfect balance. For instance, water evaporated from the ocean will be carried by air currents to the rainforest, where it will be cooled, fall to the ground, and eventually be deposited in a stream and returned to the ocean to close one loop.

Another cycle will be quite different from one aboard a spacecraft. During the day sunlight will stream in through glass panes and heat the air inside. To prevent overpressurization, a bellowslike device will allow the air room to expand. At night, with no external power from the Sun, the air will lose heat through radiation, the bellows will contract to restore pressure, and the next morning the process will start anew. In space, between Earth and Mars sunlight will be ever present, its intensity decreasing once past Venus on the outbound leg and increasing again as the Earth and Sun become closer on the trip home. Not much sunlight will be allowed into the spacecraft, whose cabin volume has no choice but to remain constant and whose pressure will therefore be regulated by allowing more or less oxygen to enter from the storage tanks.

Biosphere II will depend on photosynthesis to consume the crew's carbon dioxide and to replenish the oxygen supply. This O_2–CO_2 cycle can be considered as one closed loop or a whole array of them, for there will be hundreds of different species of plants, all working hard. Some of them are expected to die—which is consistent with the trial-and-error nature of the experiment—but the survivors are expected to fill any gaps. Some human intervention will be required, such as pruning trees and other farming activities. The Biospherians even plan to have ten coffee trees, enough for an occasional cup, and a vineyard to produce a token amount of wine.

In addition to the plants, animals such as goats, chickens, fish, and hummingbirds will be included. The goats will provide milk, the chickens eggs and meat, and the hummingbirds will be used to pollinate plants. The fish, an African species called tilapia, are involved in an interesting cycle. Waste water from the tilapia can be used to nourish hydroponic plants such as kale, strawberries, onions, and rice. Then the water flows into a drum, where bacteria convert ammonia from the fish waste into fertilizer to be fed to ferns. Surplus ferns, in turn, can be used as food for the tilapia, vegetarians that reproduce about once a month.

The Biospherians hope to eat eight ounces each of chicken and fish per week, plus two eggs and four cups of goat's milk per day. The bulk of their diet will come from cereals and legumes, such as beans and peas. They will also enjoy assorted fruits and vegetables. All in all it should be a very healthy diet low in fat and high in complex carbohydrates. They expect to consume about 2,400 calories per day. Unlike on a Mars trip, their farming ventures will be somewhat season related despite being in a sealed container. Wheat, for example, is expected to grow better in the winter, sorghum in the summer.

The organizers of Biosphere II think that their work is applicable to a Moon base or a Mars voyage. But they stress that they will also learn a great deal applicable to life on Earth, such as how to have cleaner cities, techniques for transplanting various ecosystems from one part of Earth to another, and better ways to manage wildlife refuges and national parks. The fundamental point is that Earth itself is a single biological system; by creating a second one, we learn more about the first.

To me Biosphere II seems a grandiose, wildly complicated experiment. Parts of it—tilapia to ferns, for example—I don't think will ever find their way into a spacecraft. But it's a beginning, if an unconventional one. Saving weight by closing the life-support loops is crucial when considering Mars. Recycling oxygen and water will be mandatory. Food recycling may be too complicated. Biosphere II should provide some answers, even if parts of it seem irrelevant to spaceflight. As my friend, former astronaut Rusty Schweickart, has put it, "In my view, there will never be something like this that is put together on the ground and launched bodily into space. What will go into space is the knowledge and experience gained in a self-contained biosphere like this. . . . What they learn will be extremely important."

I hope NASA listens to what the Biospherians have to say. NASA is doing its own studies, but they have a low priority and are proceeding at a snail's pace. NASA's "strategic plan" for a controlled ecological life-support system projects a series of milestones, progressing from plant growth and radiation studies to a zero-gravity crop productivity facility, a Mars-rated prototype, and then finally an operational

Mars system. The only problem is that the plan forecasts a completion date of 2025, and NASA plans have been known to slip. Life support may indeed be the "long pole in the tent"—the term NASA people give to the one item holding up the whole show.

CHAPTER

IX

POWER

IN THE past our capability in aviation and spaceflight has generally been defined by the power available: engine horsepower in the early airplanes and later pounds of thrust for jet and rocket engines. New engines usually lead to advanced flying machines, not vice versa. The matter of power, or propulsion as it is more accurately called, is a complex one when considering Mars.

The early rocket pioneers, such as Wernher von Braun, thought that no breakthroughs in propulsion would be required for a Mars voyage, just bigger and more efficient versions of the liquid chemical rockets they were accustomed to using in the V-2s of World War II or in the intercontinental ballistic missiles that followed. Of course the original rocket, probably developed by the Chinese more than a millennium ago, used solid fuel, and solids are still used today when simplicity is favored over higher performance. Since all the fuel of a solid rocket is stored inside its combustion chamber, no fancy tanks, pumps, or valves are required as with a liquid. But unlike liquid rockets, solids cannot be turned on and off. Once ignited, they burn until depleted or destroyed. For manned spaceflight their only application in a primary role has been in the two behemoths that lift the space shuttle off the ground. It was a leak in one of these that caused the *Challenger* disaster in 1986, killing seven crew members.

Von Braun's liquid rockets burned kerosene and oxygen, and that is what was used to put John Glenn into orbit. Later, more sophisticated fuels and oxidizers, such as unsymmetrical dimethyl hydrazine and nitrogen tetroxide, were introduced. The Apollo Moon rocket, the giant Saturn V, went back to kerosene/oxygen for the first stage but used hydrogen instead of kerosene for the two upper stages. Hydrogen is more efficient, and it is used today as fuel for the space shuttle's three main engines.

The efficiency of a rocket motor is measured by a simple formula that compares what goes into the motor with what comes out. The output is thrust, generally measured in pounds. The input is fuel and oxidizer, flowing at a certain rate, and is measured in pounds per second. If output is divided by input—that is, thrust (pounds) divided by propellant flow (pounds per second)—the pounds cancel each other and we are left with seconds as the unit of merit for the motor. This number is called *specific impulse*, or *Isp* in engineering shorthand. The higher the Isp the better, as far as propulsive efficiency is concerned. The old kerosene/ oxygen combination produces an Isp of 300 seconds.

Hydrogen/oxygen yields 390 seconds, a notable improvement.

Of course other factors must sometimes be considered. The Titan ICBM that powered the Gemini spacecraft had an Isp lower than that of its predecessor, the Atlas. And yet the Titan was selected because its propellants (hydrazine and nitrogen tetroxide) have two big advantages over kerosene and liquid oxygen: the Titan's fuel and oxidizer could be stored at room temperature, unlike liquid oxygen, and they were hypergolic, meaning that they burned spontaneously when brought into contact with each other, requiring no ignition source. The simplicity of these two factors was considered more important than the slight loss in Isp. A similar decision was made in favor of hypergolics to lift the Apollo lunar module from the surface of the Moon. At the other end of the scale, some exotic propellants have been rejected because they are too toxic, temperamental, or expensive, even though they produce high Isp's. Highly corrosive fluorine, for example, yields an Isp of more than 400 seconds when burned in combination with hydrogen, yet no one considers its use seriously. Research continues in the field of exotic fuels, including concepts for "exciting" molecules by bombarding them with beams of high-energy materials. In this way Isp's of 1,000 seconds may eventually be produced.

But to achieve a *really* high Isp we must abandon chemical rockets and move into other realms. Nuclear engines have been discussed for years but have not yet progressed beyond ground tests. One historical oddity is that in the same stirring 1961 speech in which President John F. Kennedy told Congress, "I believe we should go to the Moon," he added a plea for the development of a nuclear rocket. Despite this presidential impetus, nuclear propulsion has so far failed to overcome a series of technical and political hurdles. DO WE WANT CHERNOBYLS IN THE SKY? asks a headline in the *Washington Post*. Yet the attraction of nuclear power is simple: a pound of nuclear fuel contains more than 10 million times the potential energy of a pound of liquid hydrogen and oxygen! Extracting that energy safely is the problem.

A nuclear reactor produces heat, but that is just the beginning. Heat alone does not produce thrust. To generate thrust, matter must be expelled from a rocket's nozzle. The faster the particles are ejected the better—i.e., the higher the Isp. In one popular nuclear design, hydrogen is used as the propellant. It is not burned as in a chemical rocket, but is heated intensely by passing it through a uranium reactor, resulting in very high exhaust velocities and an Isp roughly twice that of a chemical rocket.

Another variation is the so-called nuclear-electric engine. In this scheme the nuclear reactor produces electricity, which is used to propel charged particles at high velocities out the back of the rocket. Isp's ten times higher than those of chemical motors are possible with nuclear-electric power.

In general, nuclear rockets have the advantage not only of a high Isp but of being able to operate for long periods of time. On the other hand, they tend to be low-thrust engines. Getting off the ground requires a huge engine, but one that need operate for only minutes. Nuclear rockets are ill suited for this. Once in space, they come into their own on interplanetary trajectories, where they might produce only a few pounds of thrust but can sustain that level for months. Another problem is that humans must be protected from the radiation produced by a nuclear reactor. This can require heavy shielding or inconvenient designs that put the crew as far away from the engine as possible. In the literature the nuclear-electric idea is sometimes called simply "electric propulsion," I think because the writers don't like to face the fact that a nuclear reactor is a necessary part of the process.

Thirty years ago the Defense Department funded something called Project Orion, one of the most powerful—and bizarre—space propulsion schemes dreamed up so far. Orion was to be a large cylindrical spacecraft, a hundred feet long and more than thirty feet in diameter. It was to be propelled by *atomic bombs* dropped overboard and detonated behind it. To prevent its being blown up by the explosions, there would be a cushioning device—an aluminum dish called a pusher plate—on the tail end of Orion. The pusher plate would have a hole in it through which a series of bombs would be expelled and then exploded some fifty feet away. When the shock wave from each burst hit the pusher plate, it would compress a gigantic shock absorber composed of gas-filled cushions and cylinders. When the absorber expanded on the rebound—sprong!—off Orion would go, considerably faster than before. Its designers calculated a Mars trip might require a couple of thousand bombs, not unlike the ones dropped on Hiroshima and Nagasaki. The crew would have a rocky ride, but supposedly well within tolerable limits. Like a nuclear rocket, Orion wouldn't have been suitable for launchpad use but only after a spaceship was well on its way.

Nuclear contamination from Orion would have been intense. Not so with the normal operation of nuclear engines, but they cause almost as much apprehension as a series of planned explosions. The reason, apart from Three Mile Island–type hysteria, centers on what might happen if a spacecraft carrying nuclear material blew up, as *Challenger* did, and radioactive debris rained back to Earth. There are probably adequate technical solutions to this problem, but the political problem remains.

Beyond nuclear propulsion, there are other fascinating ideas that can produce even higher Isp's, at least in theory. For more than half a century physicists have known about antimatter, particles that exist as counterparts, or mirror images, of normal particles. If there is a proton, for example, there will be an antiproton. Such particles are not conjecture; they have been produced in laboratories. When a proton and an antiproton collide, the result is a brief but intense burst of energy. The heat produced can be trapped in a metallic core through which a propellant gas is circulated. More complex designs might use magnets to hold antiparticles, and the heat from their collisions would accelerate plasma through a nozzle. The Isp of these schemes could vary from 1,000 to 20,000 seconds. Thrust levels of several hundred thousand pounds could result from just a minute amount of antimatter.

Then there is the Sun. Its energy is constant and considerable. Solar power satellites have been proposed to beam electrical energy to Earth, and it is just one step beyond that to use solar energy to propel a spacecraft. First the Sun's heat would be used in solar electric cells to generate electricity, which in turn would drive an accelerator that would propel ions at high speeds, producing thrust. The power of such a device would decrease as a spacecraft approached

Mars (going away from the Sun), but it might be sufficient for the task.

Another idea is the solar thermal rocket, an ingenious scheme that uses concentrated sunlight to heat a propellant. In a sense it is a nuclear device, but one whose reactor—the Sun—does not have to be carried on board.

Finally there is the solar sail. The Sun spews out an amazing amount of energy, not just light and heat but also particles such as electrons. It has been estimated that the Earth intercepts only one-billionth of the energy coming from the Sun, and that if the solar flux hitting just one square yard of Earth could be totally converted to energy it would heat and light a small room. Solar energy also exerts pressure on any object in its path. Small object, small force, but if a very thin Mylar sail of extraordinarily large area (a square mile or so) could be attached to a spacecraft, we would have a celestial sailboat and could hitch a free ride through the Solar System. Again, such a system is useless for takeoff and landing but could be unfurled en route, where it would be used to accelerate a vehicle slowly over a long time.

All these various propulsion systems might someday become practical, but in the meantime only two—chemical rockets and nuclear reactors—are technologically mature enough to be serious candidates for an early Mars mission. In the case of the nuclear reactor, however, there is still the problem of how to convert its heat efficiently into thrust, not to mention the political problems associated with its use. For example, the Nuclear Test Ban Treaty of 1963 prohibits nuclear testing in the atmosphere or in space, and it is doubtful that a space engine could be properly tested underground, as bombs are. Assuming a solution to that problem, there remains the opposition of millions of people to any device that might cause radioactive debris to fall out of the sky, no matter how remote the possibility. For these reasons, as well as an ingrained reluctance to depart from proven technology, I agree with von Braun. Go to Mars—at least at first—with conventional chemical rockets. Maybe we can get their Isp up close to 500 seconds, but beyond that—wait for some breakthrough, perhaps in antimatter.

Although I do not think any breakthroughs await us in the field of chemical propellants, that is not to say that today's rockets can't be improved and don't need to be. In mounting a Mars expedition, a large fraction of the money will be spent getting the machines assembled in Earth orbit. Today it is just too expensive to put something into orbit—roughly three thousand dollars per pound. Delivery by shuttle costs more than that, and shuttle payloads are limited to around 40,000 pounds. This country needs a "big dumb" booster—something the size of the old Saturn V, which could put 250,000 pounds into orbit—but something simpler and cheaper. We need it to make space a less expensive place for reasons (military and civilian) having nothing to do with Mars, but obviously as the cost goes down Mars becomes more accessible. To achieve lower costs we can take advantage of new materials and manufacturing techniques, but more important than either is a new approach to booster design, one that emphasizes cost over performance. We need a dump truck, not an Indianapolis racer.

In discussing propulsion, I have occasionally thrown in the word *power.* According to my dictionary there are fourteen different definitions of *power*, but in the space lexicon *propulsion* applies to engines and *power* refers to the electrical power needed aboard a spacecraft—not to get from point A to point B but to keep things running along the way. A spacecraft is a complicated machine, full of moving parts. Fans whir, inverters hum, valves pop open and closed, radios crackle, heaters cycle on and off, the microwave cooks—all these gadgets require electrical power, lots of it, and it must be not only plentiful but ultrareliable.

Early spacecraft used the simplest electrical device, the storage battery. With no moving parts, a battery produces a limited supply of direct current, which can be put through an inverter for equipment requiring alternating current. But batteries are very heavy. A more advanced form of direct-current generator is the fuel cell. Remember the old high school science experiment in which an electrical current passed through water produces one glass vial of oxygen and another of hydrogen? The fuel cell does the reverse: it combines hydrogen and oxygen to produce electricity and water. Unfortunately, batteries and fuel cells must be ruled out as primary power sources for Mars because they don't last long enough.

Then there is the solar panel. Like the solar sail, it has the wonderful advantage of operating without on-board fuel. Postage stamp–size photovoltaic cells convert the Sun's energy directly into electricity. The *Skylab* space station had two large solar panels, each thirty-five feet long and containing more than thirty thousand of these silicon cells. For a Mars trip, many more than two panels would be required. Traveling farther and farther away from the Sun, the power produced would gradually be reduced by more than half. Even worse, in orbit around Mars, or on its surface, solar panels would not work at all at night, and supplemental sources of electricity would be needed.

Nuclear energy may be the answer. Generating electrical power is a much simpler process than developing a nuclear engine. Although I reject the premature use of nuclear *propulsion*, I think some type of nuclear *power* generation will become acceptable and desirable. It may take either of two forms. The first and simplest is to encase radioactive material and convert its generated heat into electricity. We did that on Apollo Moon-landing flights with an RTG, or radioisotope thermoelectric generator. Like a battery, an RTG has no moving parts. It simply gets hot, because of spontaneous decay of radioactive isotopes within its hardened case, and that heat is used—by a device called a thermoelectric couple—to generate electricity. The plutonium-filled RTGs we carried to the Moon powered instruments there long after we departed. The *Pioneer 11* unmanned spacecraft, with an RTG on board, left Earth sixteen years ago and is now beyond Pluto, sailing out into infinity. Its RTG still provides the power that enables *Pioneer 11* to report back to Earth on conditions it encounters, such as the diminishing solar wind.

A second type of nuclear device, much more complicated than the RTG, is the familiar—and feared—nuclear power plant, or reactor. In a reactor a core of uranium, U-238, is enriched with its radioactive isotope U-235. Within this mixture atoms split (a process called fission) to form new elements, such as plutonium, and heat is given off as a result. The heat is carried away by a gaseous or liquid coolant and used to generate electricity. The amount of heat is regulated by inert rods inserted into the reactor. As more rods are inserted more deeply, the fission process slows and the reactor cools. As rods are withdrawn the process speeds up, and it could "run away," as at Three Mile Island and Chernobyl, or even melt down. In that case highly radioactive debris may be released. In Earth orbit, this pollution would eventually sift down into the atmosphere. Between Earth and Mars the only harm would be to the crew, who would certainly perish.

Space nuclear reactors are not forbidden by treaty; in fact, the Soviets use them routinely to power radar reconnaissance satellites in low Earth orbit. In my opinion reactors can be made so reliable that they would be one of the least likely parts of a Mars ship to fail. The crew would certainly accept a reactor (as do nuclear submarine crews) and appreciate it for its ability to deliver large amounts of power month after month with a minimum of attention. Even after making allowances for shielding its deadly radiation, a reactor would weigh less than batteries or fuel cells or solar cells of equivalent power.

And equivalent power is the key, because the Mars spacecraft will be large and complex and will need a great deal more power than its predecessors. NASA is currently designing a permanent Earth-orbiting space station and estimates it will need 75 kilowatts of power. A Mars expedition will probably require about the same amount. That translates into more than sixty *Skylab*-size solar panels or at least half a dozen very large RTGs. Fortunately, the Department of Energy and NASA are working jointly on a 100-kilowatt nuclear electric plant for use in space, called the SP-100. It seems to offer more power for the weight required than anything else on the horizon.

After we solve the propulsion and power questions, build a big dumb booster, and figure out how to save weight by closing the oxygen and water loops, we will be well on our way to Mars. Other aspects of the flight, such as navigation and trajectory control, might seem to require breaking new ground because Mars is so far away, but we have already sent, with amazing accuracy, unmanned probes far beyond Mars—even successfully aiming through the rings of Saturn. Steering commands from Earth can be calculated and transmitted quickly. The twenty minutes it might take for such signals to reach Mars will be more of an annoyance than a serious impediment. We astronauts learned on the Mercury, Gemini, and Skylab programs how to operate autonomously for those short periods when we lost radio contact with a ground station. On Apollo we spent about fifty minutes out of every two-hour lunar orbit on the backside of the moon, totally on our own. The Apollo flights also addressed the problem of pinpoint landings on a strange planet, so that technique is well in hand, and we learned how to return to Earth at high speed—and slice precisely into the Earth's atmosphere.

There is, however, one idea not applicable to the Moon that we *can* use approaching Mars. Unlike the Moon, Mars has an atmosphere, albeit a very thin one. Changing the speed of a spacecraft by burning rocket fuel is very costly because carrying extra fuel means extra weight. It doesn't matter whether the spacecraft needs to accelerate from a launchpad or decelerate into Martian orbit, the burn still takes plenty of fuel. If we could use the Martian atmosphere to slow us, the amount of rocket fuel could be reduced and weight saved. When an Apollo command module returned from the Moon, it used atmospheric drag to slow down by keeping its blunt end forward, which also dissipated the frictional heat caused by smashing into the atmosphere. We can apply the fundamental idea of the heat shield to Mars, but because the atmosphere of Mars is so thin, the shield must be much larger, probably around one hundred feet in diameter. Engineers refer to this extrapolation of the heat shield (which on Apollo was twelve feet in diameter) as an *aeroshell*, and to the process as *aerobraking*.

When we came back from the Moon we dug into the atmosphere at a shallow angle, about 5 or 6 degrees below the horizon. The Apollo command module generated a small amount of aerodynamic lift, and we used this to our advantage by flying the craft in the direction that would keep us on target. At first the deceleration was very slight, but as the air thickened, our speed fell off until we reached the point at which we were captured by the Earth's gravity; we then unfurled three parachutes and landed in the sea. If we had come in at too steep an angle our craft would have disintegrated, as though we had hit a wall. If too shallow, we would have skipped out of the atmosphere, like a flat stone off water, and looped around in a huge, elliptical orbit that would have returned us to Earth in a month or two. A ship approaching Mars will not land on the first pass, as we did, but will use its aeroshell to decelerate to the point at which it skips back out of the atmosphere—but not until it has slowed enough to be captured by the planet's gravity. In this way it will change its trajectory from an interplanetary fly-by to a stable Mars orbit, using friction rather than fuel. Aeroshells will also be used for the return to Earth's atmosphere.

An aeroshell is more complicated than a heat shield because it must be larger yet proportionately lighter and is more complex aerodynamically. The heat shields used on Mercury, Gemini, and Apollo were symmetrical in shape.

For a Mars vehicle, the shape of the aeroshell itself will produce lift, requiring an intricate oblate design rather than a simple conical shape. A lot of work, using wind tunnels and powerful computers, must be done to design the aeroshell properly. It is within our technological grasp, but we have not yet done it.

CHAPTER

X

LIFE IN ORBIT

PROJECT APOLLO solved in miniature many of the problems facing the designer of Mars equipment. The fundamental scheme is to design a mother ship with a lander attached: the two will then separate in orbit around the destination planet. There are also profound differences. The Moon is three days away, Mars nine months. The Apollo lander served as base camp. On Mars much more elaborate equipment should be deployed, leading eventually to a permanent colony. In consideration of these differences I think two mother ships and two interdependent landing craft should be employed. The primary reason for two round-trip vehicles is safety: if something should go critically wrong with one, an emergency transfer to the other could save lives and salvage the mission. Having two landers provides safety of a different sort, by permitting the separation of cargo and crew into more easily managed packages: two smaller, more maneuverable craft rather than a single massive one. The first will land cargo, and the crew will follow in the second.

The landers, like the Apollo lunar module, have a very specific task and will have to be designed from scratch. For example, their landing gear should be built to withstand the shock of landing on a planet whose gravity is 38 percent of that on Earth. No more, no less—a custom-tailored job. The mother ships, on the other hand, present design problems that can be solved by adapting or building upon existing space machinery. They simply have to operate in space for a long time. That time, with minor modifications, might be spent in orbit around Earth, Moon, or Mars—or in transit somewhere.

NASA is now getting serious about putting a space station into Earth orbit and keeping it there indefinitely. It has been studying the design intensively for half a dozen years and is presently issuing contracts to industry to build it. The space station, which should be in operation a couple of years before the end of the century, is designed as a laboratory to investigate making things in weightlessness and to learn more about the human body. From a manufacturing point of view, weightlessness offers some strange new possibilities. For example, it may be possible to make pharmaceutical products of the utmost purity. On Earth gravity causes sedimentation and convection currents in a heterogeneous fluid, but in space a process called electrophoresis, in which suspended particles move through a fluid in response to an electrical current, offers the promise of a superior product. It may also be possible to grow giant crystals for biological research and for applications in electronics. New metal alloys have also been mentioned. These things are possible; NASA's challenge will be to prove they can be done at down-to-earth prices. Biomedical research will focus on how to reduce bone demineralization and other harmful effects on the body caused by weightlessness.

But apart from laboratory facilities, the station (to be called *Freedom*) will be home for eight crew members. Initially NASA plans to rotate crews every ninety days, but this interval can certainly be stretched. In other words, NASA is planning to build something quite similar to a Martian mother ship for non-Martian reasons. Of course *Freedom* will lack the means of propulsion to move it out of Earth orbit, but as a long-duration habitat of approximately the right size, it should be examined in detail as a precursor to a Mars vehicle. We might also look back at *Skylab,* a station that in 1973 and 1974 kept three-man crews in Earth orbit for as long as eighty-four days. However, *Skylab* was not built from scratch as a space station but was converted from the emptied third stage of a Moon rocket, and therefore in some respects was a peculiar design. But *Freedom*'s habitat module is being designed for that purpose alone.

The space station will consist of a single-truss structure more than five hundred feet long, to which four modules are attached: three laboratories and the habitat. Assembly, supply, and maintenance of the station will be the task of the shuttle, so any component must be sized to fit inside it and weigh no more than 40,000 pounds. Assembly alone may require twenty shuttle flights. A big dumb booster, when it comes along, could do the job better, but meanwhile NASA has no choice but to design parts to fit the shuttle's sixty-by-fifteen-foot cargo bay. That means the habitat must be slightly less than fifteen feet in diameter, probably too narrow for an ideal Mars round-trip vehicle. Yet a lot of people on Earth live happily in house trailers less than fifteen feet wide. Of course, they can go out into the yard when they feel like it. The laboratory modules will be provided by the United States, the European Space Agency, and Japan. Canada will produce a mobile servicing center that will include a manipulator arm to be used in assembling the station. This joint venture will allow NASA to cut its teeth on a large-scale international project prior to the decision of how to organize a Mars expedition.

Out near each end of *Freedom*'s truss will be a group of four solar panels, totaling 25,000 square feet of surface area. These will generate the 75 kilowatts of power the station will need to sustain itself and to conduct experiments. Batteries will store the solar energy when the station is in Earth's shadow, about thirty minutes out of every ninety-minute orbit. At an altitude of 250 miles, the orbit will be inclined at 28 degrees to the equator. Later NASA intends to augment the photovoltaic output of the solar panels by using a mirror to focus sunlight on a series of pipes. Inside the pipes gas will be heated and run through a turbine that spins a generator, creating electricity. The gas is then sent back to acquire more heat from the sun, and the cycle repeats. Two of these units are expected to add 50 kilowatts of power. They of course will have to be augmented by batteries for the dark portion of each orbit, but it will be interesting to see how well they work, and what we can learn about how they might perform in the weak Martian sunlight.

From the point of view of a Mars planner, the most valuable work done aboard the station will probably be the development of a closed environmental life-support system. NASA intends to close the oxygen and water loops, so that in theory only food and nitrogen will have to be brought up periodically from Earth. The nitrogen will be required because normal air (80 percent nitrogen, 20 percent oxygen) will be used, and there will be some leakage. Atmospheric pressure aboard *Freedom* will be maintained at the sea-level pressure of 14.7 psi. On Project Apollo we used 100 percent oxygen at 5 psi; *Skylab* had 70 percent oxygen and 30 percent nitrogen at 5 psi. According to a NASA spokesman, "The decision to use an Earth-like atmosphere was made to allow life science experimenters to use valid control experiments on the ground." In other words, NASA is willing to pay the penalty of heavier structure to contain a higher pressure so that if an experimenter discovers that something works differently in space than on the ground, no one can say that the difference was due to variations in the atmosphere. For Mars, weight and complexity will be the overriding factors, and some unearthly mixture of breathing gases may be used. The simplest is 100 percent oxygen, but at higher pressure it is a fire hazard, and there may be lung damage and other physical drawbacks over a long period of time.

NASA estimates that it will save 14,500 pounds every ninety days by recycling water and oxygen. That would translate into a 100,000-pound saving for our Mars mission. The system proposed for *Freedom* would not only purify waste water but would electrolyze part of it to supply breathable oxygen. Ground tests have demonstrated the feasibility of such a system, but it remains to be seen whether it will be reliable enough to depend upon for a Mars voyage. For one thing, the purity of recycled products must be assured. Experience in water treatment plants has shown that the disinfectants used can combine with organic wastes to form hazardous chemicals. In addition to chemical toxicity, infectious diseases may also be transmitted through such a treatment process. Organisms carried on board in minute quantities may flourish within the pipes, filters, and other purification devices and become a serious menace to the crew. Cleanliness throughout a spacecraft is of the utmost importance, but the environmental-control hardware is of special concern, because of the moist fecundity within its sealed systems. *Freedom* will also provide dishwashing and laundry services, amenities as well as weight savers.

Life scientists have requested that a centrifuge, some six feet in diameter, be included on board *Freedom*. Other experimenters oppose the idea, fearful that its vibrations may jiggle and ruin such delicate processes as the slow cooling of molten crystals. For that matter, some scientists think that the crew members themselves may be the worst jigglers on board, and they wish there were some way to isolate their equipment in perfect weightlessness without some clod bumping into it at precisely the wrong moment. Be that as it may, sentiment is strong for a centrifuge, the idea being that *any* gravitational condition, from weightlessness to Earth's 1 g could be duplicated on board and its effect upon plants and animals studied. The biologists speak of *gravimorphogenesis*—the relationship between gravity and development. They say that weightlessness may have a major impact on metabolism and on cells' biological processes. Being a simple heavy-equipment operator myself, I cannot comment on the validity of such a premise, but I do think we should find out. For example, spinning the plants once a day may cause their roots to continue growing down into the soil instead of poking up into the air.

One argument raging today among the cognoscenti is whether Mars crew members might require artificial gravity en route, to counteract the harmful effects of weightlessness. It can be provided by rotating the entire spacecraft, creating a centrifugal force, but it is an untidy process from an engineering point of view, and one that would cost fuel as well. It may be possible that a centrifuge aboard *Freedom* will provide an answer to this important question. But I hope we will have decided before then; I think we can do without artificial gravity, but I will save that argument for later, after I have discussed the Soviet experience aboard their space stations.

Freedom is being designed so that new modules may be added to it as the need arises. One of its advertised uses is as a "transportation node," by which NASA means a staging or jumping-off point for other places. It would be logical, for example, to assemble, test, and fuel a Mars spacecraft at *Freedom.* To do so, some sort of assembly hangar would be needed. Six astronauts could do the work, NASA says, but only if assisted in a major way by robots and "highly automated systems with self-check and fault-tolerant capabilities." I'm not sure I like the sound of fault-tolerance. At any rate *Freedom,* if it ever gets into orbit (it has been delayed a number of times), should be a versatile facility, giving us a toehold on the twenty-first century and helping to determine just how important space may, or may not, be to life on Earth.

CHAPTER

XI

UNDER THE SEA

IF WE are impatient (I certainly am) and don't want to wait for *Freedom* to provide answers to some basic questions, are there Martian analogs here on Earth that might help? The first that comes to mind is the nuclear submarine. The vast depths of the seas have been compared by novelists and scientists to outer space: both are isolated, homogeneous (dare I use the word *boring*?) media. The U.S. Navy has given a lot of thought to, and has oceans of experience with, solving problems of confinement during cruises lasting several months. A psychologist accompanied the USS *Triton* during its historic eighty-four-day submerged circumnavigation of the globe, tracing Magellan's path. In an earlier study the navy even measured the performance of volunteers in a sealed submarine tied to a dock for two months!

Although submarines are bigger (with crews of around 150), the underwater equipment is similar in many respects to what would be used on a Mars mission: a partially closed environmental loop, a nuclear reactor, an all-important pressure hull, small water tanks, a cramped galley, tiny cabins, noisy machinery. What is important to people in this circumstance? How do they not only survive but perform at their best week after week? The navy literature is full of do's and don'ts for the ten thousand sailors typically working below the waves. Apparently there is a seven-stage pattern of adjustment: (1) depression before departure; (2) elation the first week out; then (3) an increase in sick-call visits; followed by (4) depression; (5) at the three-quarter mark, an elevated mood; (6) apprehension and depression the final week; capped by (7) "channel fever," a hypomanic state. The most worrisome symptom among these seems to be the depression that grows out of a sailor's frustration with his environment: depression produces anger, and there is no accepted way to vent it. A sense of humor helps a lot. Maybe we can put together a Mars crew of stand-up comedians ("A funny thing happened to me swinging by Venus").

During phase 3, when psychosomatic complaints seem to crest, the medical corpsman on board has a particularly difficult time diagnosing chest and abdominal pain, two of the most common complaints with serious implications. He is assisted by a computer diagnostic program, plus a system for transmitting X rays, electrocardiograms, and other pertinent data back to a naval hospital. A Mars crew will certainly include at least one physician, but even so he or she needs similar support. Medical care must not only be good, it must be *believed* to be good as well. The mind and body interact in strange ways. Even overconfidence can have harmful effects. Crew members aboard the early nuclear subs worried about working so close to radioactivity. When assured that they were properly shielded, some stopped wearing the required dosimeter. A separate effort was then required to resensitize them to wear their detectors and avoid "hot" areas of the submarine. On a Mars trip a similar overconfidence concerning radiation from solar flares could occur.

Unlike a Mars mission, on submarines medical evacuation is an option, one that is most frequently triggered by abdominal pains thought to be appendicitis. Second most common are psychiatric crises, followed by chest pains and dental problems. This is a good argument for preventive appendectomies for a Mars crew, it seems to me. Also for the medical personnel to receive extensive training in tooth extraction and other dental emergencies. Although psychiatric problems aboard nuclear submarines are statistically less prevalent than among the general population, they are difficult to handle when they do occur. One experienced officer cited a number of crises he had witnessed, including two attempted suicides and one crew member who attacked the commander with scissors. Anxiety, depression, sleep problems, and claustrophobia are more frequent problems. Proportionally, more emotional problems occur on the ballistic missile boats, which stay at sea for longer cruises than do their attack counterparts. For exceptional cases, a Mars craft should carry sedatives like Thorazine, powerful enough to incapacitate a seriously disturbed crew member.

Submariners' eyesight is curiously affected by a long patrol. Because nothing inside the sub is more than a few feet away, the eye temporarily loses its ability to focus clearly beyond that distance. Crew members first notice this when they return to land and try driving their cars away from the dock. In technical terms they have more myopia, more esotropia, less accommodative power, and poor acuity. Not a good prescription for a pilot about to land a spacecraft on Mars, where depth perception will be vital. Fortunately, a Mars mission will include some time in Mars orbit, and the astronauts will be able to study the surface from long distance. Beyond that, they may require special eye exercises en route to avoid what the navy calls "submarine syndrome."

The navy does everything it can to make life pleasant underwater. Interior decoration is considered important, with a lot of wood-grain paneling and colors like peach and sand, which are supposed to impart a feeling of openness to cramped quarters. Ceilings are usually light green or blue. Toilets and galleys use a lot of stainless steel.

Cleanliness is emphasized for both aesthetic and practical reasons. Mechanical and electrical equipment definitely is more trouble-free in a clean environment, and the crew enjoys a higher quality of life. The submarine is usually cleaned completely once a week, with all hands pitching in. Ventilation filters are kept dust-free. Lint has proved to be a problem, so all clothing and towels are prewashed to minimize it. Personal hygiene is stressed, and cleanliness of ship and crew is an essential element of unit pride.

Old diesel submarines had a serious water shortage, but water production is far superior aboard a nuclear boat. Aboard Trident-class submarines, there are no water-use restrictions. On the other types, crew members may take a "submarine shower"—they turn the water on and get wet, turn it off and lather, then rinse off. Crew members usually do not shower every day unless they have an especially grimy job, such as cooking. Water usage is about twenty gallons per person per day—a weight of about 160 pounds, far more than a space expedition could afford.

The submarine navy is famous for its good food, emphasizing quality and quantity—plenty of steak, roast beef, and lobster. Officers are generally served at tables with linen cloths, while the enlisted men eat cafeteria style. Snacks are usually available twenty-four hours a day, including soft drinks, coffee, ice cream, fruit, and cookies. Storage, including plenty of freezer space, is available for ninety days, one of the key factors in determining the length of patrols.

Alcohol is considered an infrequent luxury among submariners, and at the captain's discretion they may be allowed to "splice the mainbrace"—to mix a little ethyl alcohol with fruit juice as a reward for group performance. Cigarette smoking is allowed routinely. The British navy apparently operates in reverse, with a daily rum ration but cigarettes only as an occasional reward. So far alcohol and tobacco have been kept out of our spacecraft, although I think an occasional nip would be a good idea.

Aboard some submerged submarines the periscope is also used as a reward. Sailors call going ashore "liberty," and they refer to this as "periscope liberty." A brief glimpse of sunlight, waves, sea gulls, and perhaps an island or an iceberg goes a long way toward relieving the tedium of months of looking at the walls. On some cruises, nuclear submarines surface briefly and allow the crew on deck while the air below is freshened. The space-age equivalent would be an extravehicular jaunt in a pressure suit, but that is never done for recreational purposes. It is too dangerous and is undertaken only when necessary.

Off-duty activities are quite varied aboard the subs. There are frequent movies and other videotaped programs.

Poker is played on weekends. Music can be piped into work and sleep areas, and individual tape players are very popular. Background noise can be either good or bad. The whirring of a familiar fan, for example, can be reassuring; if it stops in the night, one is instantly awake. On the other hand, unexpected voices strike a discordant note. Privacy is prized, and each bunk is equipped with a reading light and a privacy curtain. Microfiche readers are available but they are not particularly popular because, unlike a book, they can't be taken to your berth to read. On a long space mission books will be discouraged because of their weight.

Crew members are encouraged to keep busy and most do. Often classes are conducted on a variety of topics and are well attended. Each cruise, however, does have its share of "sack rats," who alleviate boredom by sleeping whenever off duty. The normal work schedule of six hours on watch and twelve hours off ignores the body's circadian rhythm. Some crew members complain of disorientation caused by these eighteen-hour "days." No specific time for exercise is provided, but facilities are available, and weight lifting is popular. The navy appeals to vanity in promoting exercise, a quite different motivation than in space flight, where astronauts know they must exercise vigorously and regularly to prevent physical deterioration.

Like a spacecraft, a submarine is a completely self-contained unit that may be out of contact with the outside world. In the case of the submarine, communications may be interrupted because of fear of detection. On a Mars mission, the twenty-minute transmission delay, solar radio interference, and planetary position will all be factors. But both submarine and spacecraft must be able to operate with complete autonomy, at least for short periods of time.

The sub crew is organized along strict military lines, with the captain as the unquestioned authority. The captain also has his perquisites, such as a private stateroom. The more successful captains seem to relax the rules a bit while at sea, and the crew appreciates that, although almost everyone emphatically agrees that someone has to be in clear command. Group meetings are important, and training is constant. Nearly everyone is cross trained to learn a second job almost as well as his own. Training is conducted on individual and team levels. Admiral Hyman Rickover is universally credited with setting an extremely high standard of crew selection and training for the early nuclear submarines, one that the navy says endures today.

Obviously, life is not all roses on a long cruise. Family separation is the dominant hardship. Some people just don't seem to fit in, and their compatriots can be cruel in probing weaknesses or attacking unyielding facades. Minor annoyances abound because of the confinement and the peculiarities of the environment. For example, there is an occasional blowback of sewage vapors when the contents of holding tanks are discharged overboard, resulting not only in noxious odors but in an unhealthy growth of E-coli bacteria. Problems like these are mentioned at reenlistment time by sailors who hesitate to "re-up"; they refer to their craft as "seagoing sewer pipes." And submarines have an open-loop system; that is, wastes are dumped overboard. A closed-loop operation aboard a spacecraft will be much more susceptible to such unpleasant side effects.

A fundamental decision aboard spacecraft and submarines alike is the extent to which maintenance and repair will be done on board. Submarines have a list of equipment designed to be worked on at sea, called Level I. Level II jobs must be returned to port and Level III to the factory. A spacecraft doesn't have the luxury of Level II or III; instead planners provide backup, or redundant components. The trick is in knowing which ones must be duplicated and which can be repaired: it's heavier to carry two of something than it is to repair just one. And yet, our experience on Apollo was that we couldn't [illegible] nd spare parts needed for repair [illegible] to break were beyond our capabi[illegible]

Another difference between [illegible] nd a small craft in space is com[illegible] ch sailor is allowed to receive o[illegible] y-grams" during an entire nine[illegible] n-sored to make sure they don't contain information so upsetting that the sailor's job performance would disintegrate. On board all spacecraft so far, talk is pretty free. At times things are busy and messages are terse, with esoteric abbreviations to cut them even shorter. Then, during quiet intervals, long chats are not only tolerated but encouraged, to bring any problems to the surface. I think these chats will be expanded on a Mars mission, to make the months seem shorter, but suppose a crew member's child dies back on Earth? I don't know how to handle that.

Another difference involves computers. Submarine documentation fairly overflows with mistrust of the computer. There are sophisticated navigational aids aboard, but officers say they don't rely on computers in navigation or in fundamental life-support systems. "We have an automatic steering and diving system. We tried it out and it worked fine, and we turned it off and went back to using people." "You want to change the level of oxygen in a submarine, then you adjust the bleed by hand." "The computer is a fancy display. It's just that." "Where our lives depend on it, we don't use computers." Astronauts are somewhat like submariners but also different. For example, the space shuttle has the ability to land under computer control, but that has never been done. The pilots can't keep their hands off the controls. Back on Gemini we had to be ordered to allow the computer to fly the reentry instead of doing it by hand. Yet during Gemini rendezvous maneuvers we were utterly dependent on an on-board computer, as we were in making maneuvers behind the Moon on Apollo. Astronauts have been forced to accept computers in life-or-death situations; submariners have not.

Since World War II our submarine service has had an excellent safety record. The *Thresher* was lost with all hands in 1963 and the *Scorpion* in 1968, but that is all. There have been some on-board deaths, generally the result of fire or explosion. NASA lost an Apollo crew in a launchpad fire in 1967, a shuttle crew in an in-flight explosion in 1986. Compared with spacecraft, submarines have had much more exposure time—their statistical data are expressed in millions of man-days. Despite the obvious differences in size and environment, I think space designers should learn all they can from submarines, especially when planning for Mars. Perhaps they should emulate Admiral Rickover's methods, at least some of them. When it comes to crew selection, I cannot resist quoting from the admiral's testimony before Congress:

> I view with horror the day the navy is induced to put psychiatrists on board our nuclear submarines. We are doing very well without them because the men don't know they have problems. But once a psychiatrist is

> assigned, they will learn they have lots of problems. They recently reported—it was in the newspapers—that they had completed a study . . . to determine the psychological difference between sailors who had been tattooed once, sailors who had been tattooed more than once, sailors who had never been tattooed but wished they had, and sailors who didn't get tattooed and didn't want to be. As you might suspect, the sailor's love life or lack of it and his attitude toward mother and father were all deeply involved. Of course it is a very complex subject and will require more study; in other words, more money.

Rickover would have made a good astronaut.

CHAPTER

XII

ICE WORLD

MARS IS a place, not a vehicle. Landing on it and surviving there will be different from plowing through the water or zinging off into space. The Apollo lunar module was a good flying machine but a miserable place to spend the night. For Mars we need a whole new approach, with the first landing considered not an end in itself but a vanguard, to be followed by a base camp and then a permanent colony. In searching for a terrestrial analog, Biosphere II comes to mind, but it has serious limitations. The Sun is too hot in Arizona; the location is not remote enough; the two-acre, glass-enclosed miniworld is organized as an agricultural research station more than a toehold on a strange planet.

Of all the places on the surface of the Earth, probably none seems as extraterrestrial as Antarctica. It is as cold as

Mars and almost as remote. Its land area is half again as large as the United States. Ninety percent of all the ice in the world is there. The South Pole base is 10,000 feet above sea level, perched on ice nearly two miles thick. Despite all the frozen water, Antarctica receives so little rain it is classified as a desert—the world's largest. The average annual temperature at the South Pole is 60 degrees below zero Fahrenheit, although it can drop to 112 below during the winter. Exposed flesh freezes in less than thirty seconds.

During the summer more than a thousand people may be in Antarctica, but most of them leave in the fall. During the eight-month winter fewer than a hundred Americans remain at four stations. In recent years these groups have ranged in size from eight to thirty. It is these "wintering-over" crews who have experienced conditions similar to the subzero, windswept plains of Mars. For eight months they are totally isolated from outside contact: no visitors, no supplies, and only infrequent radio contact. (In fairness, medical evacuation may be possible, but the timing depends on the weather.) The primary source of danger is an accident while working outside.

About half of those wintering-over are civilian scientists and half navy support personnel. They range from eighteen to forty-five years old, younger on average than a Mars crew would be. Unlike a Mars crew their educational levels vary widely, with recent high school graduates working next to Ph.D. graybeards. At each station there are a civilian group leader and a navy commander. Winter work includes maintenance of generators, heaters, and other equipment; laboratory experiments; brief forays outside; and housekeeping, primarily kitchen duties. Generally each group is a mixture of Antarctic veterans and novices, mostly the latter. The navy is responsible for their psychological prescreening. Motivation for volunteering is split: the civilians are looking for scientific discoveries, while the navy personnel are more interested in banking their salary. There's little chance to spend it in the Antarctic winter. And because so little time can be spent outdoors, free time is abundant.

Conditions inside are mixed. Compared with a spacecraft, accommodations are commodious. Sleep chamber volumes, for instance, average five hundred cubic feet in the Antarctic as opposed to forty-eight on *Skylab* and thirty aboard nuclear submarines. There are homemade saunas and sports facilities. Still, privacy and personal space are at a premium, especially for the navy personnel, who cannot escape to a laboratory. The temperature within a room frequently ranges from zero at the floor to 85 degrees near the ceiling. Water is made from melted snow and often tastes of diesel fuel. The food is plentiful, and the galley doubles as the social center. People overindulge—some groups have averaged a per capita alcohol consumption of nine ounces of distilled liquor and 5.4 beers per day. Reading tops the list of leisure activities.

During the 1957 International Geophysical Year, the U.S. Antarctic team experienced one case of schizophrenia, which turned out to be highly disruptive. Since then a number of studies have been conducted that show a higher than normal incidence of neurotic behavior during wintering-over but very few psychotic breakdowns. Furthermore, psychiatric evaluations are judged to be relatively successful in predicting individual performance. People *are* concerned about being in this strange new alien environment and in some cases their well-being is on their minds to an excessive extent. There is no place for a hypochondriac at the South Pole or on a Mars expedition.

Sleeplessness can become a problem. Unless strict discipline is maintained in the Antarctic, the tendency during the eight months of constant darkness is to go to bed later each night and remain awake longer. People miss the normal diurnal cues, which psychologists call by their German name, *zeitgebers.* The result of zero *zeitgebers* can be irregular schedules and bungled work. Occasionally insomnia becomes acute, a condition referred to as "polar big eye."

In our trips to the Moon we tried to keep our schedules keyed to local time in Houston, our home. But frequently we could not because certain events refused to conform to this timetable. You arrive at the Moon, for example, when Sir Isaac Newton's laws dictate, regardless of whether it may be bedtime in Houston. In Earth orbit the *zeitgebers* go crazy, because in twenty-four hours an astronaut sees sixteen sunrises and sixteen sunsets. On the way to Mars, as on an Apollo flight, I guess the *zeitgebers* become more or less frozen, because it is always daytime—in the sense that you are constantly exposed to sunlight. On the other hand, if you look away from the Sun and shield your eyes, the view is of a beautiful, star-studded night. I don't know, maybe that's why psychologists don't try to translate *zeitgeber* into English. On a Mars mission I suspect Houston (or Moscow or Tokyo or Geneva) time will be the basis for setting schedules, to which the crew will rigidly adhere whenever possible. They will have enough problems without developing "Martian big eye."

In any group wintering-over, there is apt to be one man who goes native in a big way. He stops shaving and then washing, and seems to derive pleasure from "grubbing out." He is usually a scientist rather than one of the military. In tight confinement, such conduct becomes magnified in importance to his companions. Supposedly Captain Cook, on his voyages into Antarctic waters, required his men to bathe once a week and to change clothes frequently. No less is expected today, in Antarctica or in space, although with water being a far more precious commodity on a Mars voyage, cleanliness may become a real problem no matter how hard a person tries. Certainly the superclean standard of normal American life will have to be relaxed. But slovenliness has a way of escalating, as in this account from a station leader's log:

> One civilian's coffee cup had become so dirty that I threw it in the garbage can. We had soup for evening meal and he used his cup. After finishing his soup he hung his cup in the rack without cleaning it. About an hour later he found the cup and wanted to know why I had put it there. I told him why, whereupon he lost his temper and started acting like a child.

While tempers do flare frequently, the more common response to the long winter is the same as in the long submarine patrol—depression. I expect Mars will be no exception, even though those selected will have undergone much more rigorous testing and their objective will be much more clearly defined—and more important to them. Frequently in Antarctica an outburst is followed by withdrawal, or "cocooning." Another log entry:

> Cook's at it again. He's moody, definitely emotionally immature. Threw a lemon pie and cookies all over the

> galley the other day, then went to his room for a couple of days and wouldn't come out.

Some experienced hands believe that food is so important that the quality of the cook determines the success or failure of a group's winter-over experience.

Strange schisms and adhesions develop in wintering-over groups. Classical-music-loving scientists may gang up on navy youngsters who prefer country and western. Yet factions quickly band together when the group turns to the outside. The same thing happened to one of the *Skylab* crews in its relationship with Mission Control, an "us versus them" attitude. "They" don't understand "our" problems: they don't understand how hard we are working, how difficult it is when instruments freeze and stick, when equipment floats off in weightlessness, when the food lockers are labeled wrong. We are doing things right and it would be obvious if the clods got off our backs.

Some wintering-over crews have had a special diversity not found aboard submarines or *Skylab:* the presence of women. In a twenty-two-month trip to Mars and back, I would not want to be cooped up with only men. *Skylab* crews readily admit that sex was very much on their minds as the months rolled by. But only in Antarctica, as far as I know, have small, isolated mixed groups been studied.

Women were first permitted at U.S. Antarctic stations in 1969, resulting in headlines like POWDERPUFF EXPLORERS TO INVADE THE SOUTH POLE. Then a nun and a female biologist wintered-over. The presence of women eases some stresses and creates others. On one hand, it seems to elevate standards of civility, compared with the raucous, "animal house" conduct of all-male groups. Most experts feel stability and productivity are increased by mixing the sexes. On the other hand, promiscuity has been known to disrupt Antarctic stations. Unattached men seem to accept pairing, especially if it occurs early, although if the station leader is involved, hostility is apt to result because the other males feel he has an unfair advantage. Beyond simple pairing, difficulties abound. As *Skylab* astronaut Bill Pogue put it, "People will accept a sequestered couple, but there are problems with bed hopping, where there is a hope for sharing."

Husband-and-wife teams seem a good solution, but in the case of our Mars trip, it may be extraordinarily difficult to cover all required disciplines and skills with a group of four married couples. Maybe a marriage certificate is an unnecessary frill. Maybe the most highly qualified people should be selected, irrespective of sex, and to hell with the raging hormones.

Homosexuality is another question. If, as some people claim, 10 percent of men are homosexuals, then statistically those picking a Mars crew will be faced with some highly qualified homosexual candidates. I would not pick them. I think enough interpersonal problems will develop among a totally heterosexual crew, and introducing an element of homosexuality could only serve to make matters worse. I guess the same principle applies to lesbians.

People have suggested that sleeping accommodations aboard the space station *Freedom* be designed like some motels, with doors connecting each two sleeping compartments that can be opened provided they are unlocked on both sides. This simple test of consent sounds reasonable to me. As a naval psychiatrist remarked of the Antarctic, "Having a woman around can make the condition much more endurable. They are different and they can remind you of home." They also tend to be smaller and use less oxygen, water, and food: a weighty argument in their favor—if any more reasons are needed.

The exploration of Antarctica, unlike that of other continents, has taken place in the name of science. The programs of the United States are organized and directed by the National Science Foundation, with support from the Department of Defense. It was not until the International Geophysical Year that extensive efforts were made to establish bases in Antarctica. The United States proposed an international treaty to assure scientific access to the continent. The treaty, in effect since 1961, will be reviewed in 1991. Seven nations (Argentina, Australia, Chile, France, New Zealand, Norway, and the United Kingdom) do have territorial claims, but neither superpower recognizes their validity. The treaty declares that the continent shall be used for peaceful purposes and that member nations must support scientific exploration there. The treaty is unusual in that it provides for the right of access and inspection to all bases. In that regard it could serve as a model for Mars, where scientific exploration should be paramount. At present Mars is covered by the United Nations Treaty on the Peaceful Uses of Outer Space, signed by the United States and the Soviet Union in 1967. It stipulates that outer space should benefit all mankind, and that no nation may claim any celestial body. But the Antarctic treaty is a bit more specific.

If Antarctica is a good legal model, it is a bad one in ecological terms. The main U.S. base at McMurdo Station, located on Ross Island, is an example of how *not* to treat pristine surroundings. Trash is burned outdoors, and the local waters are so polluted by chemicals and other wastes that most flora and fauna have been killed off. Mars deserves better treatment. In the summertime, McMurdo uses 30,000 gallons of water a day, a luxury unthinkable on Mars.

Some of the science being performed today in Antarctica is applicable to the exploration of Mars. One theory is that Mars once harbored life that has since become extinct. Microbes in Antarctica have been found that survive the harshest conditions by staying in rock crevices. There they extract minerals, including iron, from the stone in a strange process unlike any used by other terrestrial organisms. The rock changes color where they are living, or have lived; it loses its brown and acquires a whitish patina. As life disappeared on Mars, so the theory goes, rock interstices might have been the last refuge of microorganisms. When people get to Mars they should know what to look for.

Antarctic lakes may also be valuable in predicting conditions on Mars. There are no lakes on Mars today, but the likelihood is that there were plenty of them at one point in the planet's evolution. Primitive life on Earth is generally believed to have begun about 4 billion years ago, within 500 million years of its formation. Mars is thought to have been warm and wet in its first 500 million years, although it has since cooled off and lost atmosphere. Scientists are studying ice-covered Antarctic lakes and finding green slush on their surface and dome-shaped sediment deposits on their bottom. Certain features on Mars seem to resemble stromatolites, thick carpets of fossilized microorganisms. Again, Antarctic research may be useful in selecting landing sites on Mars. Parts of Antarctica certainly look like Mars, with volcanic landforms and dry-rock river valleys.

It would be interesting to organize a forty-day stay in a

habitat designed specifically to simulate the first Martian landing. That is, a crew of four, supported by the equivalent of two Mars landing modules, including a nuclear power plant and a vehicle for exploring the surrounding territory. The crew could go through a "postlanding" phase, getting themselves organized and their equipment deployed; then an exploratory period, with strict water rationing; and finally, "prelaunch" preparations. Of course such a simulation could take place much more conveniently in Houston, Texas, or elsewhere, but there is merit in using Antarctica. It has snowstorms similar in some consequences to Martian dust storms. It has low temperatures and pressures, high winds, and similar topography. But most of all it has isolation. To "In vino, veritas," I would add "In isolation, truth."

CHAPTER

XIII

LISTENING TO TSIOLKOVSKY

THE SPACE age began in 1957 with *Sputnik*. The first man into space was Yuri Gagarin, the first woman Valentina Tereshkova, the first space walker Aleksei Leonov. While we Americans debate the design details of *Freedom,* the space station we hope to put up, the Soviets make improvements to *Mir,* orbiting since 1986, not to mention a series of *Salyut* stations before that. The Soviet Union is a space-faring nation second to none, and has had its sights fixed on Mars for a long time.

Early in the twentieth century a small-town Russian schoolteacher, Konstantin Tsiolkovsky, wrote seminal treatises on spaceflight, establishing the theoretical foundations for the hardware that eventually followed. As early as 1903, Tsiolkovsky envisioned multistage rockets fueled with liquid hydrogen. He variously described his feelings about leaving planet Earth, the most quoted being to the effect that the Earth is a cradle, but mankind cannot endure in its cradle forever. His work was continued by other pioneers such as Fridrikh Tsander, who labored in obscurity during the 1930s in an organization called the Moscow Group for the Study of Reactive Motion. But it was not until after World War II—after von Braun's V-2 rockets provided a glimpse of real possibilities—that space research flourished. Both the United States and the Soviet Union captured some German rocket scientists, and of course we got von Braun himself. But it would be very wrong to say that the Soviet scientific and military establishments merely provided support to the captured German brainpower. As Walter McDougall has so thoroughly explained in his Pulitzer Prize–winning book, *The Heavens and the Earth,* the Soviets were culturally and bureaucratically intrigued by Tsiolkovsky's ideas, and during the 1920s and 1930s built a technological and organizational base that allowed them to proceed swiftly once the horrible devastation of World War II had been put behind them. German help certainly accelerated the launch of *Sputnik,* but probably not by much. Most of the credit should go to Sergei Korolev, who until his death in 1966 was referred to only as the "Chief Designer."

On October 4, 1957, the space age began with Korolev's launch of *Sputnik 1,* Earth's first artificial satellite. Weighing 184 pounds, and looking like a cannonball trailing four short wires, *Sputnik 1* circled once every ninety-six minutes, emitting beeps for the startled world to hear. Only a month later the Chief Designer's team followed with *Sputnik 2,* containing a small dog named Laika (Barker). *Sputnik 2* weighed over a thousand pounds and kept Laika alive for ten days before running out of oxygen.

The two *Sputnik*s were remarkable achievements and caused near-panic in the United States, which scrambled to launch its own satellite. Our first attempt, dubbed *Flopnik*, was destroyed on the launchpad. It rose a few inches, was engulfed by flame, and collapsed, and its payload was tossed onto the concrete a few yards away, where it lay emitting plaintive bleats. Four months after *Sputnik 1*, we succeeded with *Explorer I*, which discovered the Van Allen radiation belts and was a great scientific success. But *Explorer I* weighed only thirty pounds, and its rocket was not nearly powerful enough to reach the next step—putting a man in space. That milestone was passed on April 12, 1961, with the single orbit of Yuri Gagarin in a Vostok spacecraft. Our Mercury program responded with suborbital flights by Alan Shepard and Gus Grissom and three orbits by John Glenn. The space race was on. Our Gemini program produced advances the Soviets did not match with the Voskhod, and by the time they had worked the bugs out of their Soyuz late in 1968, we were orbiting the Moon with *Apollo 8*. They stayed in Earth orbit and organized for a space station.

Mars has been a strong, consistent theme for the Soviets for fifty years. The Soviets, perhaps more than others, have felt an affinity for the Red Planet. "Even back in the thirties, when Tsiolkovsky was alive, that was our dream," according to rocket pioneer Igor A. Merkulov. Since the sixties the Soviets have used "Forward to Mars" as the slogan for their space program. In the meantime they have done just about everything we have done except fly people to the Moon. For years they claimed they never wanted to do that, although a cosmonaut did tell me privately in 1967 that he had been assigned as commander of the first circumlunar flight—one that never took place. Once, in 1975, an Apollo and a Soyuz joined up in Earth orbit, and three astronauts and two cosmonauts exchanged food and expressions of eternal good will. Both sides have suffered losses along the way: a launchpad fire and a shuttle explosion killed ten Americans, and four Soviets perished in a parachute malfunction and a spacecraft depressurization. All along the Soviets have specialized in long-duration flight; they are far ahead of us in that regard, having logged more than three times as many man-years of orbital time. They have never ventured beyond low Earth orbit with people, or beyond Mars with automated probes.

The Soviets have had a philosophy different from ours in regard to advancing their space capability. We tend to go in spurts and to take large technological leaps. They are more conservative and generally improve existing products and techniques rather than plunge into new ones. Their basic booster, the type that put Yuri Gagarin into orbit, has made more than a thousand flights. Their Vostok, Voskhod, and Soyuz rockets were variations of the same design, not different from one another as Mercury, Gemini, and Apollo were. They have not been as advanced as we in electronics, especially miniaturization. The Apollo crew that joined the Soyuz in orbit reported that the Soviet craft seemed quite crude compared to the sleek Apollo command module, designed for a lunar round-trip.

The Soviet planetary probes have had a spotty record. They have tended to concentrate on Venus, leaving Mars to us for the time being. I think part of the reason is that Venus is more accessible (three or four months away), and therefore to them the most logical sequence seemed to be Venus-Mars. They knew we were going to send Vikings to Mars and did not want to compete in that arena. Furthermore, their early experience with Mars probes was terrible, one failure after another; they licked their wounds by retreating to Venus. Just a partial listing of the accomplishments of their Venera series: 1961, a flyby; 1965, surface impact; 1967, atmospheric research; 1972, soil analysis; 1975, photographic data; 1983, radar images. They have so much information about Venus that their scientists have not yet digested it all. Their only successes with Mars were a flyby in 1961 and a soft lander in 1971 that apparently succumbed to a dust storm and stopped transmitting data after twenty seconds. They have studied our Viking results eagerly and concluded that our scientists are unduly pessimistic about the possibility of life on Mars. Many Soviets believe, or profess to believe, that Viking didn't prove much one way or the other, and that life somewhere on Mars is still a distinct possibility. They say, "We haven't been there and therefore we don't have to be as pessimistic as you."

Today the Soviets have more powerful launchers, more sophisticated sensors and electronics, and greater confidence in their use. I believe that Mars will be the focus of their efforts over the next dozen years. For me, a look at the current Soviet approach began in July 1988, when I was preparing an article on Mars for *National Geographic*. Working through Novosti, the Soviet news agency, I invited myself to the Soviet Union. *Glasnost* was in early bloom, although Moscow did not seem any more hospitable than on several previous visits. However, I was surprised to be allowed to proceed thirteen hundred miles southeast of Moscow to the remote province of Kazakhstan. Situated there in a bleak desert east of the Aral Sea is the long-secret Baikonur Cosmodrome—sometimes referred to as Tyuratam, the name of a nearby town. Sprawling for many miles across the Steppes, Baikonur is a much-expanded version of Cape Canaveral and the location of all manned Soviet launches. I was there, along with a small group of foreign journalists and photographers, to witness the launch of a complicated probe to Phobos, the larger of Mars's two moons. The booster rocket was a Proton, a twenty-year-old workhorse powerful enough to speed its 13,700-pound payload to escape velocity. We spent an afternoon touring hangars and launchpads and then, a couple of hours after dark, watched the Proton liftoff from a distance of two miles, about as close to one of those things as you want to get.

All in all I was impressed by Baikonur and the people I saw working there. Like all Soviet buildings, even new ones, its hangars and administrative centers seemed to be of the shoddiest construction, clad in cracked, discolored concrete. But inside, the equipment looked well maintained if not quite up to the spit-and-polish standards of equivalent U.S. facilities. There was an air of casual, almost nonchalant competence, and several old hands were introduced who had spent their entire careers at Baikonur launching rockets. When the carefully calculated T minus zero arrived, the Proton simply *went*—no fancy NASA-style countdown over the loudspeakers. It was a perfect launch, and *Phobos 1* was on its way to Mars. It was followed five days later, on July 12, 1988, by *Phobos 2*, a backup vehicle with similar capabilities.

Their mission was an intricate and fascinating one: a two-hundred-day, 120-million-mile cruise and then a rendezvous with tiny Phobos. Several weeks after launch, chances for success were cut in half by an egregious blunder. Someone in Mission Control sent an erroneous command to *Phobos 1*, shutting off its guidance sensor. The spacecraft was then no longer aligned with the Sun and began to tumble. Its solar panels, which needed to be pointed at the Sun, lost power—and *Phobos 1* slowly died.

Phobos 2 continued on and reached Mars in January 1989. To try to intercept Phobos directly would have been more difficult than hitting a bullet with a bullet, so the strategy was to sneak up on the target. The first step was to decelerate into a highly elliptical orbit that was later tilted to coincide with Mars's equator. The ellipse was then reduced to a smaller, circular orbit. By late March *Phobos 2* had taken a number of photographs of Phobos and the surface of Mars and was ready to commence its rendezvous with Phobos. At this point disaster overtook *Phobos 2*. The spacecraft changed its antenna position in response to a ground command, but it failed to return to the alignment necessary to acquire new signals from Earth. With no way to communicate with it, Moscow's Mission Control personnel could only watch helplessly as *Phobos 2* continued its solitary, but now useless, orbit around Mars. While some useful scientific data were obtained, the two Phobos failures were serious blows to the Soviet planetary program. The two probes were ingenious designs whose operation would have been fascinating to observe.

Phobos 2, for example, carried two landing devices, one designed to anchor itself to Phobos and the second to hop over its surface. After dropping to a height of 150 feet over its target, *Phobos 2* was designed to survey the surface for fifteen to twenty minutes. It carried a laser whose beam would have blasted the surface, causing vaporized ions to ricochet up into a test chamber that would have studied the chemical and physical properties of the soil. The surface of Phobos would also have been scanned by radar and television.

The first of the two landers was to be anchored to the surface by a harpoonlike device that had plunged a yard deep into the soil. Anchoring would have been necessary because in the extremely weak gravitational field of Phobos—one-thousandth of Earth's—the machine might have rebounded from its initial landing and caromed back into space.

The hopper looked like an inverted kettle with thin metal legs sticking out below it. The legs were designed so that by rotating them, the hopper could right itself if overturned. Then the legs would snap together, propelling the machine some one hundred feet at a time, unhampered by gravity. It was expected to hop perhaps ten times, measuring surface conditions such as acceleration and magnetism as it went.

These ambitious Soviet plans must now be added to a long string of earlier Martian failures. The Soviet interest in the planet remains undiminished, but their attempts to visit it robotically have consistently ended in frustration. Their probe designs have not included sufficient safeguards against mechanical malfunctions or human errors.

Closer to home, in Earth orbit, the Soviets have enjoyed much greater success. They have two new machines, a shuttle named *Buran* (Snowstorm) and the Energiya rocket, a big dumb booster. The Energiya is in the same class as the Saturn V rocket that propelled us to the Moon and put *Skylab* into orbit. It looks different, though; the Saturn V was tall—365 feet—and skinny, whereas the Energiya is short and squat. It is about 200 feet tall and 65 feet across at the base. It can put 220,000 pounds into low Earth orbit, just slightly less than the Saturn V's payload. The reason the Energiya is so compact is that instead of having all its stages piled on top of each other, it uses strap-on boosters clustered around the base of the main rocket. Unlike our shuttle, which uses solid rocket strap-ons, the Energiya's are liquid fueled (kerosene and oxygen). At liftoff a total of twenty engines are blazing away, compared with the Saturn V's five. The strap-ons are separated when empty, fall to the ground, and may or may not be used again, depending on which Soviet account you believe. The second stage is a large central core, using liquid hydrogen as a fuel for its four engines. Third-stage engines are attached to the payload. To describe the Energiya as a "big dumb booster" implies that it is inexpensive, and I don't know if that is so, but if our experts believe they can drastically reduce launch costs by building such a machine, I expect the same holds true in the Soviet Union. At any rate the Energiya is the only big dumb booster around today. The Saturn V flew its last in 1972, and its dies have been destroyed.

While Energiya is quite different from the Saturn, *Buran* is about as close to the U.S. shuttle as carbon paper allows. There are two explanations for this. The first, and more charitable, is that each is designed to solve the same problem, using similar technologies and materials. Modern jet airliners of the same size, for example, look pretty much the same: a sleek fuselage, wings, and a vertical fin swept back for optimum subsonic cruise. Only the placement of the engines and horizontal stabilizer distinguishes one from another. A more prevalent view is that once again the Soviet designers have benefited by technology transfer—simply waiting until the West develops something and then borrowing it. Our shuttle, although it carries military payloads, was designed and built by civilians, and the literature is unclassified and available to anyone.

If the two shuttles were parked side by side, the only difference a layman might notice is that the three rocket motors protruding from the tail of our shuttle are missing on the Soviet version. Our shuttle is attached to an external liquid fuel tank, which in turn has two solid rocket boosters hanging from its sides. At liftoff the two solids ignite, as do the three engines in the shuttle's tail, burning fuel from the external tank. When the solid boosters and the external tank are empty, they fall away, leaving the three motors to repeat their task on the next flight. *Buran* uses the Energiya as a booster, and all the motors are attached to it. They may be reused, but because of the damage they incur when they parachute to earth, I doubt it. Probably parts of them can be salvaged.

Once the Energiya rocket and *Buran* are separated, the latter becomes a more graceful glider than our shuttle because it doesn't have to cope with the weight and drag of three engines hanging out its tail end. Even more important is the fact that *Buran* uses the Energiya as a booster instead of two solids and an external tank. Without the shuttle attached, our solids and tank are useless, but the Soviet system can launch either manned or unmanned, depending on whether a big dumb booster or a sophisticated manned vehicle is required for a particular mission. For example, if the Soviets have a payload that's too large for *Buran*'s cargo bay (virtually the same size as ours: fifteen by sixty feet), they can launch it on the Energiya. Thus they have a versatility that we don't, having put all our eggs in the shuttle basket.

Buran was launched in November 1988, more than seven years after John Young and Bob Crippen piloted the shuttle *Columbia* into orbit. *Buran* flew unmanned, an indication of the different approach the two countries take to matters of safety, reliability, and public relations. The conservative Soviets were not willing to risk human lives on the first, or even second, *Buran* flight. But they did trust their electronics sufficiently to conduct a completely automated flight, including an airplane-type landing. NASA gave a great deal of thought to an unmanned first shuttle flight but couldn't swallow the notion of the great beast booming in over densely populated southern California, *unmanned,* for a landing at Edwards Air Force Base. It was too scary. On the other hand, NASA felt confident enough in their untested design to put Young and Crippen aboard. Of course part of the difference is geography: unlike Edwards, Baikonur is in the middle of nowhere, so no sizable population would be at risk during a bungled approach or landing. But beyond that, the Soviets generally take a cautious, step-by-step approach, and their analysis indicated that *Buran* should fly unmanned at first. The cosmonauts wanted to fly it from the beginning, just as Young wanted to fly *Columbia,* but they were overruled by the program managers. Yet the Soviets had sufficient confidence in the Energiya to entrust *Buran* to it on only its second flight. That is a departure from past procedure, and indicates a certain maturity of hardware and management. The Soviets are proud of their newfound capability, the heart of which is computer technology, and they discuss applying it to allied fields such as automated aircraft landings, weather forecasting, and even control of highway traffic.

Western experts believe the Soviets are building at least

three shuttles in addition to *Buran*, but they are stumped as to exactly what the Soviets will do with them. For most orbital jobs it is simpler, safer, and cheaper to use the Proton or the Energiya. About all the Soviets have said is that they intend their shuttles to deliver people to space stations, but they certainly don't need four for that alone. In fact the Soviets indicate a launch rate for their shuttles of only four per year, adding to the mystery. Just as scientists in this country have criticized our shuttle program for not being responsive to their needs, so in the spirit of *glasnost* Soviet voices of dissent can be heard. According to Roald Sagdeev, the former head of the Soviet Institute for Space Research, *Buran*'s first flight was "an outstanding technological achievement" of "absolutely no scientific value." This is an example, perhaps, of politics overtaking technology. When the Soviet shuttle program was begun in 1976—twelve years before its first flight—Communist Party chief Leonid Brezhnev may have envisioned a future quite different from that seen now by Gorbachev and Sagdeev.

Certainly nothing like *Buran* or *Columbia* will ever venture out of Earth orbit. They were not built to do so, and they can carry only thirty days' worth of supplies. But the Energiya is equipped, according to former cosmonaut General Gherman Titov, "with the most powerful of modern liquid-fuel rocket engines . . . [and] can launch a probe to either Mars or Venus with a mass of up to twenty-eight tons." It could also be used to assemble parts—big chunks—of a manned caravan to Mars.

There is absolutely no doubt about Soviet intentions regarding Mars. Yuri Gagarin used to speak of a Mars mission, pointing out that oxygen and water would be recycled and that automated landing systems would be required.

Cosmonaut Aleksei Leonov, first to venture outside a vehicle in space, was quoted in Japan in 1970 to the effect that the Soviets wanted to send a man to Mars within ten years. In 1976 cosmonaut Georgi Beregovoi wrote of expeditions to Mars. In 1982 I listened to the Soviet science attaché in Washington predict a manned trip within ten to fifteen years. None of these statements was fleshed out with details, but international cooperation has been a strong theme in the Soviet space literature, and therefore it is not surprising for them to consider Mars as an international objective.

But it was surprising—to me at least—to pick up my copy of the *Washington Post* one morning in December 1987 and read a long article by Roald Sagdeev. At that time Sagdeev was the closest thing to a NASA administrator that the Soviet system had; his comments were specific and quite detailed:

> . . . The Soviet Union and the United States need to work together on important scientific projects at the frontiers of knowledge. As a start, we should go to Mars together.
>
> We should begin carefully, with unmanned missions. First, by 1994, we should cooperate in landing a mobile, unmanned space vehicle on Mars. Later, by 1998, we should cooperate on a mission to bring samples of the Martian planet back to Earth. That is truly a challenge for the century. If all goes well with these missions, we could try to cooperate in landing men on Mars, maybe by the year 2001.
>
> We should be realistic. If Americans are worried about transferring sensitive military technology to the Soviet Union, we should find ways to work cooperatively, short of fully integrated missions. For example,

> we could each send payloads to Mars that would be launched separately from Earth but work together on Mars. . . . This would continue the sort of cooperation we showed in the 1975 Apollo-Soyuz docking. . . .
>
> The cost of these Martian missions would be manageable—far below what our two countries now spend annually on nuclear arms. . . . A major, manned expedition to Mars would cost $50 to $100 billion, assuming the dollar will be stabilized.
>
> This joint effort would not only broaden the dialogue that is now under way between our two countries but also bring a new dimension to it. If our dialogue remains based on arms control, then only military aspects will be involved and we will have difficulty establishing the language of mutual understanding. Something stronger should exist.

There is no doubt that Sagdeev speaks with authority: he is a deputy in the Soviet Congress and has been officially declared "a hero of Socialist labor," and he is a close friend and adviser of General Secretary Mikhail Gorbachev. He is also candid and outspoken.

In an article entitled "Science and Perestroika," he writes, "French, German, and American science is vibrant with new ideas, while Soviet science is stultifying." In a speech to the Soviet Academy of Sciences he suggested that an honor intended for him be given instead to Andrei Sakharov, and he noted dryly that he had been invited more often to testify before committees of the U.S. Congress than at "any kind of hearings at the Supreme Soviet in my own country." In an interview with Louis Friedman, executive director of the Planetary Society, Sagdeev stated that *perestroika,* or the restructuring of the Soviet system, is necessary because "the country was badly managed for seventy years."

But in the same interview, Sagdeev seems to retreat from the forceful approach he took in the *Washington Post* article, saying instead: "I don't know how realistic it would be to talk about the manned flight to Mars right now, but I'm sure if both nations decided on such a goal, technology would allow us to do it the first ten or fifteen years of the next century." That is quite different from "maybe by the year 2001." I don't mean to nitpick; rather, I'm trying to suggest as accurately as I can the current thinking in the Soviet Union—and Roald Sagdeev is, in my opinion, able to state space policy as accurately as anyone except Gorbachev himself. Gorbachev has agreed to Mars in principle, saying, "I will offer . . . cooperation in the organization of a joint flight to Mars. That would be worthy of the American and the Soviet peoples." One final point: Would the Soviet Union go it alone, wondered Louis Friedman during the interview, if the cooperative approach wasn't working fast enough? Sagdeev: "In principle, yes. . . . [But] I would be terribly unhappy if we had to do it."

What might the Soviet approach to Mars be? Generally a three-part program is proposed. First, a mission in 1994 to orbit Mars, release one or more balloons into its atmosphere, deploy small meteorological stations on the planet's surface, and drop penetrators to gain more knowledge of surface and subsurface characteristics. This flight would be followed by an unmanned landing at a spot selected using the 1994 data; a rover would cruise the surface, taking samples that would be returned to Earth. If all went well, a manned expedition would follow, either directly or after an unmanned dress rehearsal, with the first humans stepping on Mars around 2010. Naturally these plans—and dates—are fluid, and factors such as the failure of *Phobos 1* and *Phobos 2* may significantly alter them.

In 1988 cosmonaut Vladimir Solovyev, a veteran of 352 days in Earth orbit, attended a technical conference in Los Angeles and presented his view of a manned Mars mission. Energiya boosters, perhaps ten of them, would launch pieces of the Mars spacecraft that would be assembled at a Soviet Earth-orbiting space station. A crew of ten would take nine months to reach Mars and spend three months on the surface. En route, artificial gravity would be provided. One module would be dedicated entirely to medicine, to "treat the most serious cases of sickness or injury in space." Two physicians might be needed, said Solovyev, who stressed that the medical aspect is the greatest concern of Soviet planners. He emphasized psychological stress and "the problems of compatibility of crew members, the problems of maintaining the relationships with your family . . . and also with the flight directors." Solovyev said he expected the first landing to take place after the year 2000.

Usually when cosmonauts speak, they echo the official Soviet line, unlike their American counterparts, who are apt to "wing" it. *Pravda* is even more likely to present the government's view, and it published a long article on Mars a couple of months after Solovyev's comments. It was quite technical, with discussion of trajectories, trade-offs of fuel versus trip time, the safety advantages of having two vehicles, the use of nuclear propulsion to save weight, protection from radiation and meteorites, flying by Venus, aerobraking, and the quarantine of the crew upon return. The article estimated that using chemical rockets, fifteen Energiya launches would be required, a number that could perhaps be reduced to five by using nuclear-electric engines. It argued that artificial gravity would not be necessary.

We have a similar difference of opinion in this country concerning artificial gravity. In regard to nuclear power and propulsion, however, the Soviets seem to be more bullish on its development than do our people. I think we are more sensitized to public opinion after the accident at Three Mile Island: new nuclear applications are viewed with the greatest suspicion. Despite the Chernobyl disaster, the Soviet Union is pressing on with various designs. It has tested in space a new nuclear reactor called Topaz; researchers are impressed by its ability to provide substantial electrical power for a long time despite its small size. An earlier type of reactor powers the military's Radar Ocean Reconnaissance satellites that the Soviets use to track the movements of the U.S. Navy. When one of these satellites nears the end of its life, the nuclear reactor separates from the satellite and goes into a higher, "graveyard" orbit, where it should remain for centuries. Occasionally this feature malfunctions and the reactor enters the atmosphere and disintegrates. In 1978 one strewed radioactive debris across remote northwestern Canada. No one was injured but the debris might have been lethal had an unsuspecting hiker found a piece and kept it in his or her backpack for a couple of days. According to Soviet documents translated by them into English, current planning for a trip to Mars features two round-trip vehicles called NEJ 1 and NEJ 2, NEJ meaning Nuclear Electric Jet.

In April 1987 an agreement on cooperation in space science was signed by Secretary of State George Shultz and Foreign Minister Eduard Shevardnadze. It was amended during the Reagan-Gorbachev Moscow summit of 1988. The agreement establishes five working groups covering space biology and medicine, Solar System exploration, astronomy and astrophysics, solar-terrestrial physics, and Earth sciences. Sixteen cooperative projects have been agreed upon, but they are all small, and the closest to a Mars mission is a joint study to identify the planet's most promising landing site. Thus it is that the eager Soviet suitor is being stalled, for reasons to be discussed later in this book.

CHAPTER

XIV

366 DAYS

THE SOVIET UNION is now practicing for Mars in Earth orbit. More than submarines or Antarctica, this is the place—and for the time being at least, it belongs to the Soviets. "We have no competitors here," said Academician Kotelnikov, celebrating Cosmonautics Day in 1986.

The Soviets have been putting up space stations since 1971. A series called *Salyut* has kept crews of two or three up for a cumulative total of more than nine man-years. Now the latest in the series, *Salyut 7,* has been supplemented by *Mir* (Peace), a more elaborate station. By the end of 1988, the Soviet Union had accumulated nearly four times as many man-hours in Earth orbit as had the United States. In their methodical stair-step way, the Soviets have increased mission length as follows (in days): 23, 62, 96, 140, 175, 185, 211,

237, 326, 366. At this writing they are attempting to stretch the interval even further. With the capability to rotate crews at will, the only sensible reason I know for such durations is to assure themselves the human body can handle weightlessness all the way to Mars and back. Along the way they have learned a lot, some of it confirming what common sense would tell us, but also other lessons that are not so obvious.

The basic *Mir* module is about fifty-five feet long and fourteen feet in diameter. It can hold up to six people. It has six docking ports so that it can be easily expanded. The Soviets have plans to add to *Mir* gradually. At the present time, only one module is attached to it, an astrophysics lab called Kvant. Next in line is an airlock, to be followed by modules for materials processing, remote sensing, and medical research.

Speculation in the West is that *Mir* can be modified for a trip to Mars, and that the Soviets might attempt at least a flyby of the planet fairly soon, perhaps in 1992, to commemorate the seventy-fifth anniversary of the Bolshevik revolution. Judging from their literature and from talks with a half dozen cosmonauts, I do not think such a possibility is likely. If the Soviets go it alone to Mars, I don't think it will happen until well after the turn of the century. Propulsion is the first problem. They genuinely seem to feel they need to develop a nuclear-electric engine. Second is consumables; although *Mir* stays up indefinitely, it is dependent upon *Progress* for supplies at fairly frequent intervals. Third is the crew and the Soviet assessment of how its members might fare, physically and mentally.

It is to this last point that most of my attention has been devoted. I do not understand Russian, and reading English translations can be maddening at times. Some of the stuff I read makes no sense to me, but I can't tell if it's because the Soviets are behind or ahead of our experts or because the translator just blew it. Example: the cosmonauts receive "moral-political" training. Isn't that an oxymoron? Example: "Doctors were concerned that Lyakhov, a solidly built person, might on his first encounter with weightlessness experience problems. Therefore, they concentrated the greater part of his training on tuning his vestibular system. Ryumin's training concentrated on his cardiovascular system." As far as I know, whether or not a person is "solidly built" has absolutely nothing to do with the functioning of his inner ear. Yet clearly this passage indicates the Soviets are tailoring physical and medical training to individual attributes, and I don't understand why or how. Maddening!

To start with the simple things, the average age of the cosmonauts is forty-five. An odd-numbered crew seems to work together better than an even number—or at least it's been learned that three is better than two over the long haul. Food is very important, especially its variety and seasoning. (Garlic overpowers the ventilating system.) A little bit of vodka is allowed on board (hurrah!). Cleanliness is a must: The space station walls are of washable leather, a nice Russian touch, and they are wiped weekly with a disinfectant. The cosmonauts take showers every ten days, a cumbersome procedure that seems to consume most of the day. Restful colors are used for sleep chambers; the *Salyut 7* wardroom has one apple green wall and one beige. Music can be piped in. (Supposedly experiments have shown that classical music—unlike more discordant types—has a positive effect on the cardiovascular system.)

Exercise is much, much more important than here on

Earth. Cosmonauts have exercised as much as four hours a day. On some flights there have been two sessions per day, but lately it has been one two-hour workout daily, with an occasional day off. Generally the crew that has been most faithful in executing the regimen is in the best shape after landing. For example, Yuri Romanenko was in better condition in 1988 after his 326-day flight than he had been a decade earlier after 96 days when "my legs felt like they had been made of lead." The difference is attributable mostly to exercise.

Immediately after returning to Earth the long-duration crews are unable to stand. They are helped from their spacecraft and ceremoniously hauled around like nabobs in sedan chairs. Generally they regain their legs in a matter of hours and are walking well within a couple of days. It is interesting to consider their condition in the context of landing on Mars rather than Earth. Some planners are quite pessimistic, pointing out that arriving unable to walk is not exactly a felicitous condition for any explorer. But I don't think it will be all that bad. In the reduced gravity of Mars, they may spring back almost immediately from a year of weightlessness. Furthermore they don't have to plunge out onto the surface right away, as we did on Apollo missions. They might be better off studying their new surroundings and planning their forays for a couple of days while they regain their strength and balance.

In addition to the heavy emphasis on exercise, the Soviets have tried to preserve muscle tone and circulatory functioning with mechanical devices. First is the "penguin" suit, a pair of overalls with rubber braces embedded in them. The rubber bands force the wearer to stoop and to use muscles to straighten up, somewhat like fighting against gravity. The

Valles Marineris stretches over two thousand miles along the Martian equator and reaches a depth of nearly five miles. Named for the U.S. Mariner spacecraft that spotted it in 1971, this huge canyon system dwarfs any similar feature on Earth. Huge ancient river channels flowed from the east of Valles Marineris into a basin called Acidalia Planitia, the dark area at the north in this picture. The *Viking I* landing site is located in Chryse Planitia, south of Acidalia Planitia. Three of the Tharsis volcanoes (dark red spots), each about fifteen miles high, are visible to the west.

A photograph of Chryse Planitia taken by *Viking I*. After *Viking I* arrived in Mars orbit, scientists studied the surface for nearly three weeks before deciding on this rock-strewn plain as the best landing site.

Looking due west on July 26, 1976, the *Viking I* lander took this self-portrait.

A frozen dry valley in Antarctica. Studying evidence of life in the desolation of Antarctica is providing clues as to how we can best approach exploratory missions to Mars.

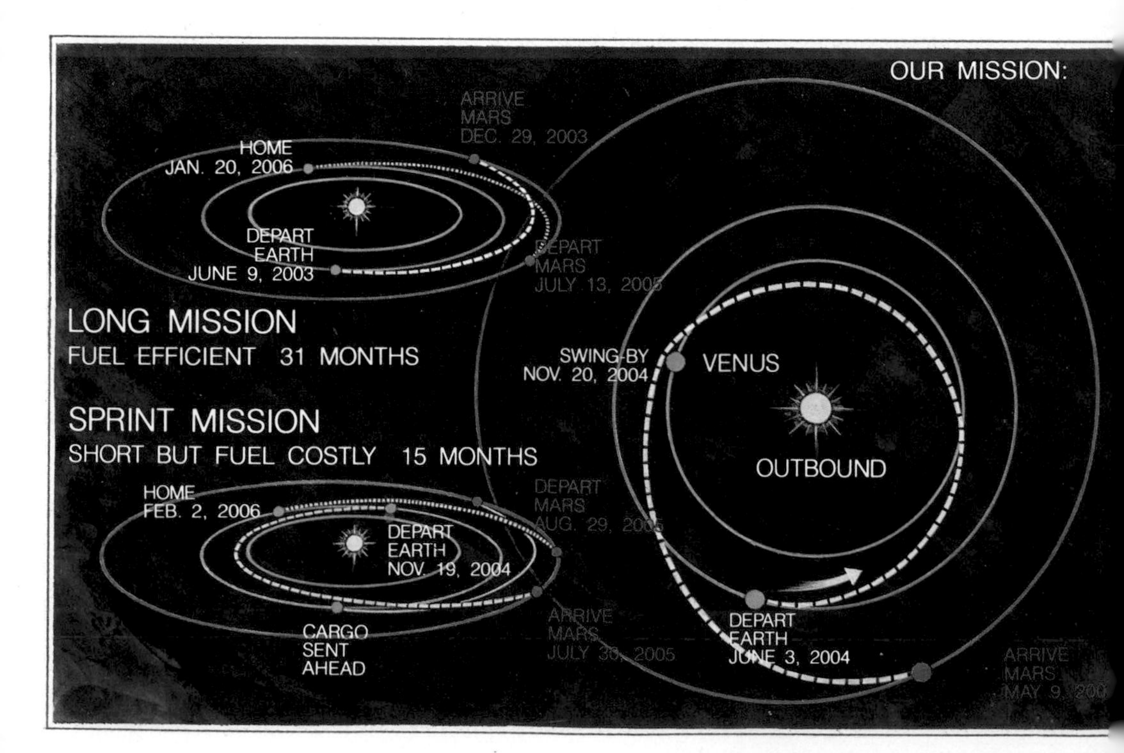

The most fuel-efficient route to Mars is a direct trajectory (top left) taking less than seven months each way but requiring a stay on Mars of over a year.

The fastest route is a sprint mission, which reduces the round trip to about fifteen months but greatly increases the fuel load.

Our mission combines advantages of both scenarios. By using the gravitational field of Venus to "sling" the spacecraft toward its destination, the mission achieves necessary velocity while conserving fuel. The landing crew spends about forty days on the surface of Mars before the nine-month trip home.

Nearing Mars (far right), the Soviet and U.S. ships separate, then "aerobrake" to reduce speed (1) and enter an elliptical orbit around the planet (2). Next the ships jettison their aeroshells (3), recouple (4), and target the landing site. After separating again, the crew's landing vehicle (5) and the unmanned supply vehicle (6) descend to the planet's surface. After forty days of exploration, the crew launches the landing vehicle's built-in ascent stage (7) to rendezvous with their fellow crew members in the orbiting mother ships. The craft blast out of Martian orbit separately (8), then redock for the journey home (9).

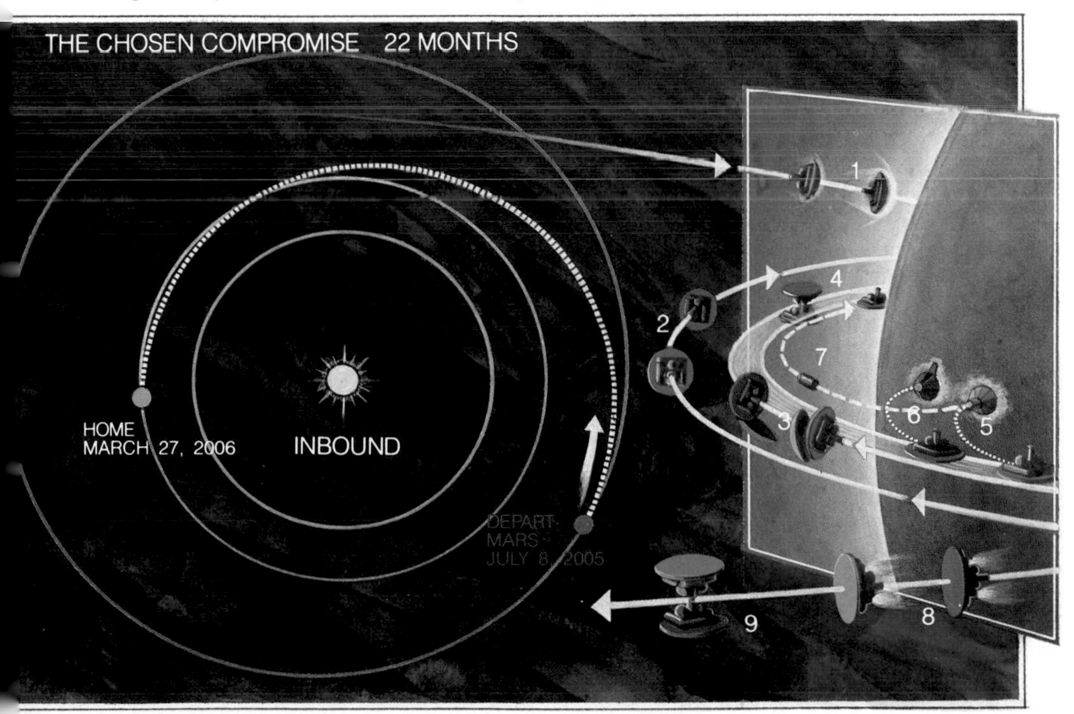

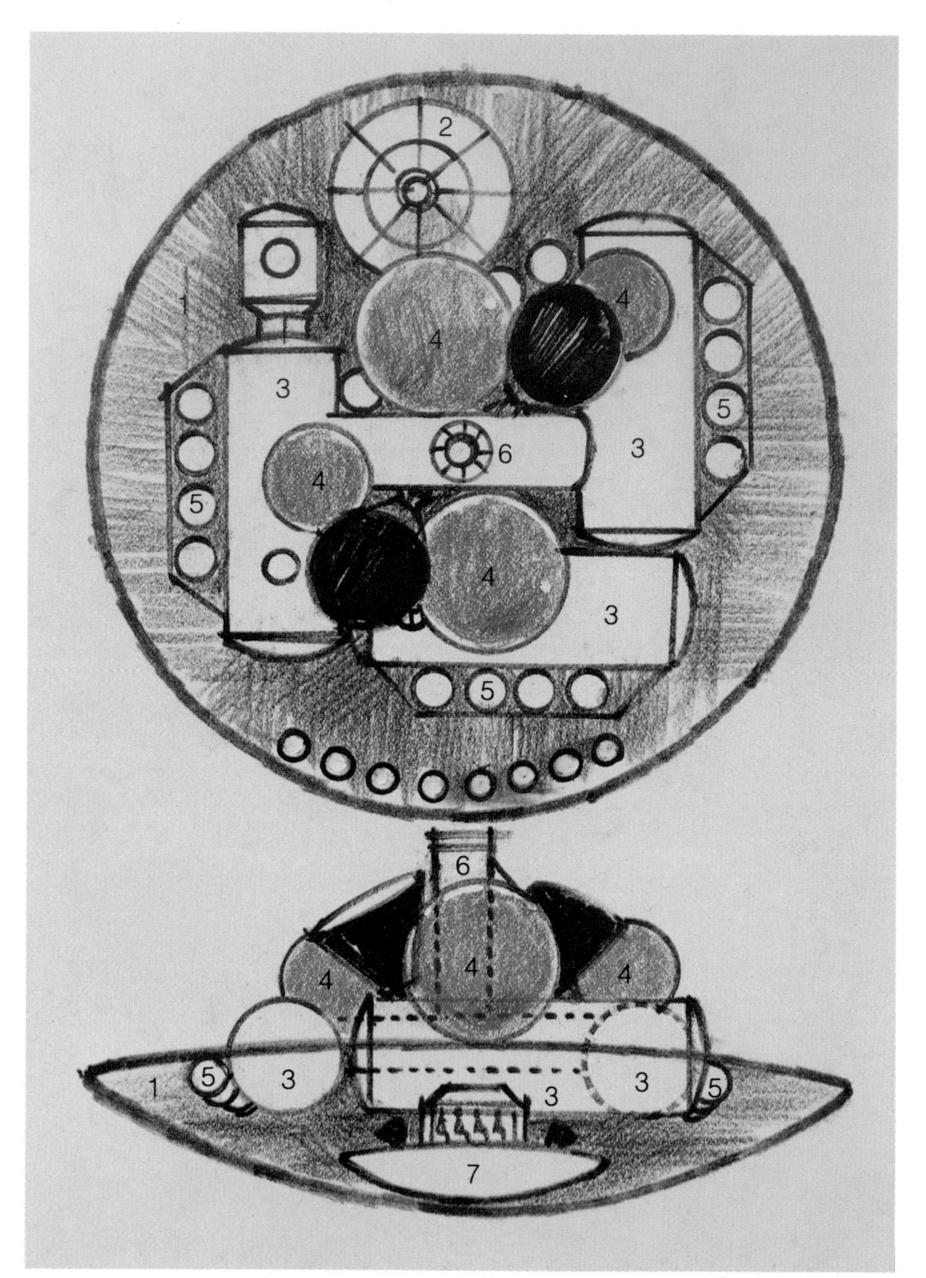

One of two round-trip modules, this spacecraft includes an aeroshell (1), a Mars landing vehicle (2), habitation modules where the crew lives and works (3), fuel tanks (4), water tanks (5), a docking tunnel (6), and an "earth-capture" vehicle (7) with its own aeroshell.

Riding atop a pair of escape rocket stages, the U.S. round-trip module departs Earth. Seen at left is the assembly sequence, during which unmanned heavy-lift rockets delivered components of the Mars craft to space station *Freedom*, where crews fitted them together. A similar operation conducted by the Soviet Union produced a second round-trip vehicle, which will travel to Mars docked with its U.S. counterpart.

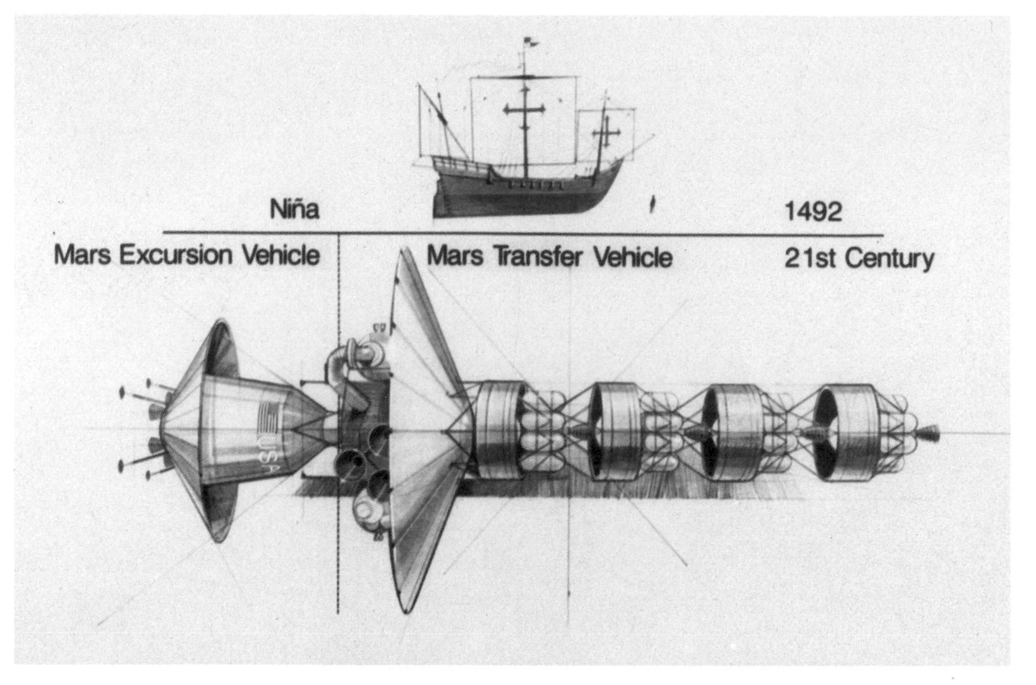

A scale comparison of a fifteenth-century ocean-exploring ship and its twenty-first-century counterpart designed for an expedition to Mars.

En route, the crew must face problems ranging from deadly bursts of solar radiation to the physical and mental stress of months of tight confinement, with Earth a mere speck of light in their windows.

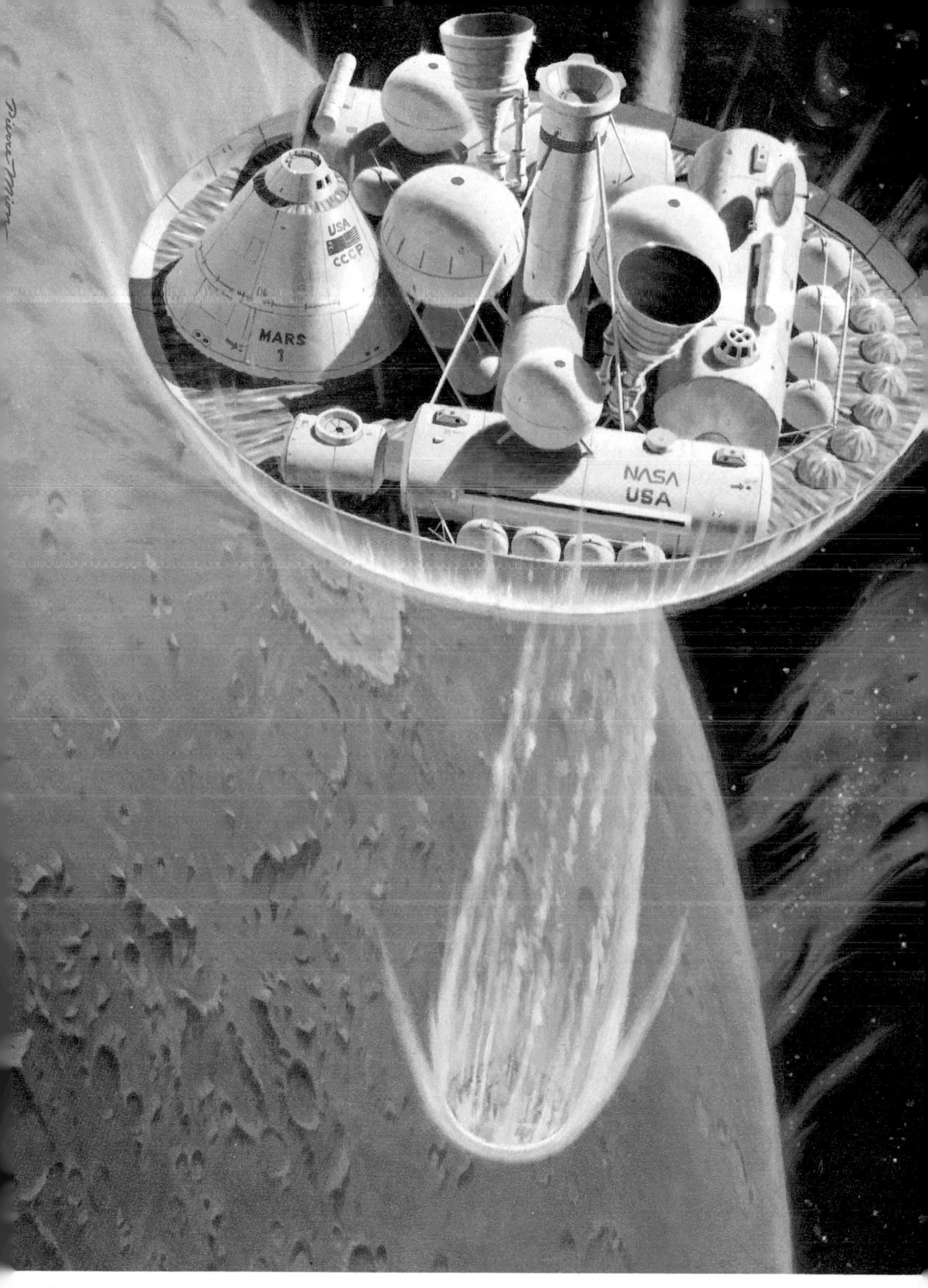

Plunging into the Martian atmosphere with their aeroshells forward, the Soviet and U.S. ships use friction rather than fuel to reduce speed and enter an elliptical orbit around the planet.

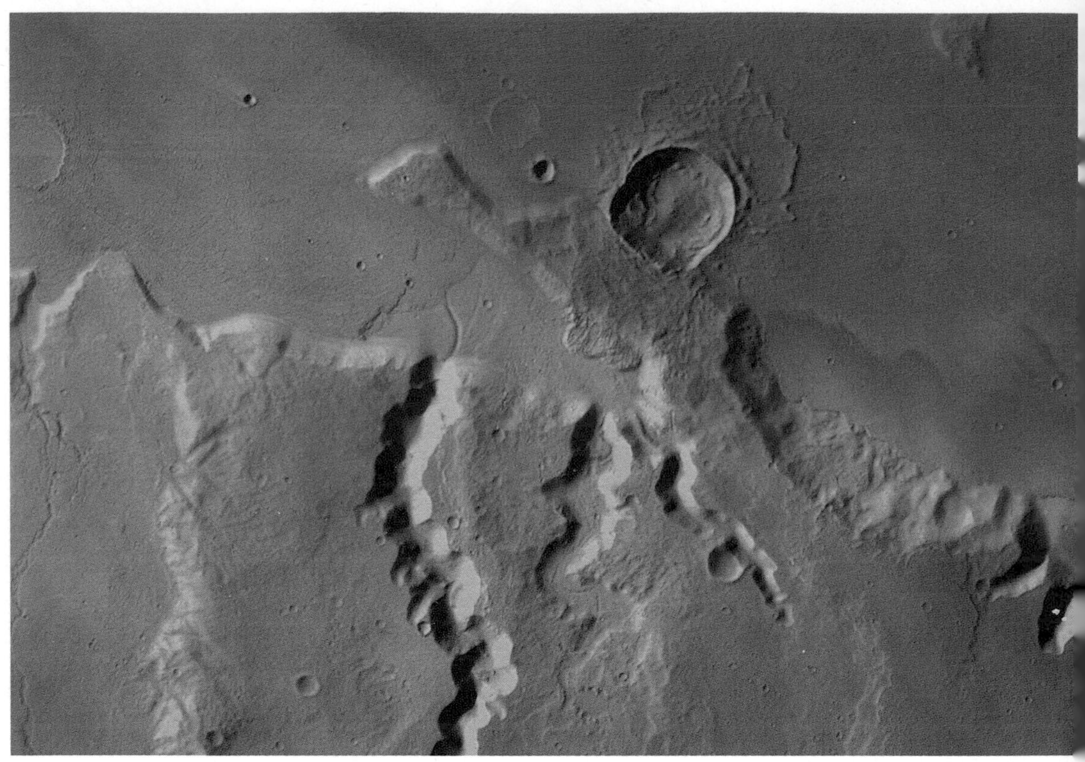

Our landing site, in the eastern Mangala Vallis, where ancient river channels, lava flows, and impact craters are clustered.

A collage of potential activities at a Martian outpost, limited only by human ingenuity. Surveying and prospecting; producing oxygen; growing plants; sending out rovers, balloons, and robotic planes: colonists will be busy establishing first a foothold and then a comfortable habitat.

second is a trouserlike garment, sealed around the waist, that can create a negative pressure on the lower body, forcing blood toward the legs and simulating the pull of gravity. Third is a leg bracelet fitted to the thigh to control excess blood flow to the upper body. Finally, a couple of days before landing, cosmonauts drink extra water and take salt in an attempt to restore a terrestrial fluid balance. All these ideas apparently help, but it is the long hours on treadmill and stationary bicycle that are crucial.

According to cosmonaut Valeri Ryumin, "In a profession like mine, where there is constant stress, a cosmonaut must have control over his feelings and emotions." This attitude, coupled with the usual Soviet penchant for secrecy, makes it difficult to gauge precisely how the long-duration crews have performed, but clearly there have been problems. According to Georgi Beregovoi, who is in charge of cosmonaut crew training, despite diligent preflight testing to assure crew compatibility, it takes about thirty days for signs of hostility and irritability to emerge. The preflight training period is the best time to work out incipient problems, and at least one cosmonaut blames in-flight disagreements on the fact that preflight togetherness was curtailed: "Our mutual relationship was simply not solved. . . . We had to deal with it in space." According to U.S. experts, shouting matches with one another and the ground do occur. On one flight, a fracas broke out at the time a blood sample was scheduled—and it *was* conducted, one cosmonaut cutting his companion's finger to the bone. The wound then became infected. On another long-term mission, a cosmonaut became irritable and troublesome as a result of excessive doses of sleeping pills. He complained about his work schedule and the difficult living conditions. Some cosmonauts worry about their health, cosmonaut Ryumin in particular: "My worst fear . . . [is] a possible appendicitis attack. And the other: that I may get a toothache requiring dental help. One night aboard the station, I dreamed I was having a toothache and I woke up in a cold sweat. And one tooth was indeed sensitive. But by morning the toothache was gone." Usually flight crews report only minor ailments such as colds, headaches, fatigue, insomnia, and eye irritation. When hostility does develop, usually it takes the form of deliberately hiding information when talking to the ground and becoming irritated at questions deemed unnecessary. The Soviets have used voice stress analysis to judge what is going on with a crew falling into this uncommunicative pattern. Electronic equipment assists in measuring and recording the level and tempo of speech, timbre of voice change, stuttering, and slips of the tongue.

Only one flight has been terminated prematurely because of human factors. In 1985, aboard *Salyut 7*, Vladimir Vasyutin became ill. According to a crewmate's diary Vasyutin was tense, a bundle of nerves. He developed a high fever and retired to his sleeping bag. After long conferences with the ground he and his two companions returned to Earth, and he was hospitalized in Moscow. No word was released except that he had suffered a severe inflammation. According to Western experts it might have been pneumonia or a prostate infection. A couple of years later, cosmonaut Alexander Laveikin was replaced after six months aboard *Mir* because of heart irregularities during a space walk. I wonder what medical treatment these two received back on Earth, and whether it could be duplicated in the limited facilities on board a Mars craft.

Fatigue has also taken its toll. Toward the end of his 326 days, Romanenko's workday had to be reduced to four and one-half hours. Cosmonaut Valentin Lebedev wrote in his diary, "The most difficult thing about this flight is keeping calm . . . because pent up fatigue could generate serious friction." He also wrote, "I am apprehensive about myself; whether I will be able to live and work so long with my colleague; whether I will be able to keep my composure and self-control." On some flights, such as Romanenko's, the fatigue got progressively worse. On others it was cyclical and seemed to respond to schedule changes.

The personnel in Mission Control must walk a fine line between under- and overscheduling. If underemployed, the crew becomes bored and fatigued; if overworked, tension and fatigue result. Fatigue dominates both ends of the spectrum, and Mission Control must be sensitive to the condition of individual crew members and the phase of their mission.

This became abundantly clear during our Skylab program when the ground transferred the busy schedule of an aggressive, experienced trio to their replacements before the rookies got adjusted. It takes about thirty days in space for a crew to hit its stride, to learn the little tricks that weightlessness plays and how to overcome them. Like tenderfoot Boy Scouts on their first outing, new crew members lose items of equipment and waste time looking for them. Anything not tied down floats off, usually drifting toward the intake screen of the ventilating system. In time anchoring one's body at a workstation becomes second nature, but there is a lot of flailing about that wastes time and energy in the beginning. According to Yuri Romanenko, the spaceman of the future should have "a bald head, to avoid haircuts . . . big arms . . . six would be better . . . slim legs . . . or just one . . . with grips, to keep steady."

In the case of the *Skylab* crew, their fatigue caused them to make mistakes and put them even further behind. The crew commander, Jerry Carr, reported during the flight, "I think you could tell by our voices that we were very, very frustrated. . . . No matter how hard we tried and how tired we got, we just couldn't catch up with the flight plan. And it was a very, very demoralizing thing to have happen to us." His companion Ed Gibson added, "I personally have found the time since we've been up here to be nothing but a thirty-three-day fire drill." Cosmonauts have expressed similar sentiments and compared themselves to squirrels on a wheel.

A Mars flight will be somewhat different in that there will be few scientific experiments en route. The majority of the equipment carried will be to support exploration of the planet itself. Therefore, unlike *Skylab,* which conducted extensive studies of the Earth and Sun, the challenge to Mission Control on a Mars mission will be to overcome the monotony of the months in space. The Soviets are good at this kind of support. Once a week cosmonauts are allowed to talk to their families on the radio. This contact becomes more important as time goes by. Videotapes are also sent up with the supplies aboard *Progress.* One cosmonaut played a tape of his daughter's birthday celebration over and over. "I would put it on when I felt particularly homesick. You watch, get engrossed in it, and it seems like you're with your family." Cosmonauts also hold extensive conversations with celebrities such as athletes and entertainers. Tapes of Earth sounds—birds chirping, the wind in the willows—are provided. There is even a special room in Mission Control reserved for a psychological support group that orchestrates these activities.

One suggestion for a spaceship of the future is a "psychological relief room" similar to experimental facilities in factories on Earth. In such a compartment, cosmonauts or astronauts would be surrounded by large screens upon which would be projected images of, says the Soviet literature, "a flowering meadow, a birch grove, or a pond." Background music begins slowly, accompanied by the singing of birds. "After three and one-half to four minutes, the music becomes bolder and a stimulating, arousing light is turned on the wall panel. At the end of the session the room lights are turned on and invigorating music sounds . . . after these sessions the workers are in an improved state of mind . . . the overall state of the central nervous system is improved. . . . This method of psychological relief can improve work productivity up to 17 percent while at the same time reducing traumatism." I doubt that our Mars ship could afford such a facility; in fact I am inclined to put on my Mickey Mouse ears while listening to such proposals. It goes against my test-pilot ethos. But as I have indicated before, Mars is different, and I would not disparage any idea that could help keep the crew healthy and happy.

In many ways the Soviet approach to crew training is almost identical to NASA's. It begins with a heavy dose of classroom work and moves on to devices such as centrifuges and underwater tanks. Simulators are vital, and both cosmonauts and astronauts spend many hours rehearsing in them. After individuals have mastered the basics they are assigned to crews and continue their team training, focusing on the problems they expect to encounter on their particular flight. The training is intended to fulfill two functions: to produce individuals who are competent and prepared, and to make sure the crew works together as a harmonious, well-coordinated unit.

The Soviets throw in an extra: they scare the hell out of candidates. Their favorite stress producer is the parachute. They say that parachuting develops self-confidence, discipline, and steadiness during an unexpected or emergency situation. Day and night jumps are taken from different altitudes. Every cosmonaut makes at least a hundred parachute jumps while performing tasks that become successively more difficult. For example, he may be required to carry on a radio conversation, identifying locations on the ground, *before* opening his parachute. The Soviet fondness for parachuting can be traced back to Yuri Gagarin's day. The early Vostoks were not considered suitable for landing: the cosmonauts bailed out.

Astronauts are not required to undergo parachute training. NASA prefers to pick people who have already scared the hell out of themselves many times in airplanes—i.e., test pilots. My own experience with parachutes is limited to one forced ejection from a burning jet fighter and one voluntary free-fall from twelve thousand feet. There certainly *is* a big difference between a crisis in a simulator and the real-world experience of falling through the sky. Despite that, however, I don't think parachute training is necessary if a candidate has a strong aviation background. If he or she does not—for instance, if a physician were being considered for a Mars flight—perhaps parachuting or some other stress inducer would be valuable in predicting who will panic and who remain calm in the event of an in-flight emergency.

Cosmonaut training has been complicated in some cases by the presence of foreign cosmonauts. The foreigners spend one and a half to two years in training compared with much longer periods for the Soviets, from which I conclude that they are like the mission specialists on our shuttle flights—capable of some specialized tasks but not trained in all the tasks aboard a Soyuz or aboard *Mir.* The training of the foreigners includes Russian language lessons, physical conditioning, and spaceflight theory. As far as I can discern the Russians have gotten along well with all the foreigners, although their flight duration has been limited to about a week. The Soviet literature mentions that members of international crews should have similar personalities and ideas, a "union of like-minded people." To me that means they should be blue-ribbon Communists. If capitalists like our astronauts are mixed in, I don't know whether politics would be a divisive topic or not. Since *glasnost,* probably not. Most all the two dozen cosmonauts I have met seem like people I could get along with, although we have not been locked up together for months. As I used to tell John Young before *Gemini 10,* I was happy I was making my first spaceflight with him, but I wanted to fly so badly I would have gone up with a kangaroo. For a couple of days it doesn't matter, and that is the sum of our experience so far—the brief Apollo-Soyuz flight of 1975. On that one everyone got along together famously, conversing in what they called Ruston (Russian plus Houston). Translators and technical personnel did have trouble deciphering parts of their conversations. On a Mars trip language could definitely be a problem, not to mention cultural differences. In that regard the Soviets have been as accommodating to their guests as possible. For example, they provided little touches of home, such as mango juice for the Indian, who was encouraged to practice his yoga exercises in flight. Ethnic foods were provided for the visitors, and the Russian crew seemed to enjoy the variety. Maybe they were just tired of their own diet, which to my taste could be rather odd. For example, a typical breakfast menu is pork with sweet pepper, Russian cheese, honeycake, and prunes. Then for lunch it's jellied beef tongue, cherry juice with pulp, and praline candies. For dinner, ham, borscht with smoked ingredients, beef with mashed potatoes, rye bread, cookies with cheese. Wait! We're not finished: they eat four times a day. Supper: cottage cheese with nuts, assorted meats, enriched wheat bread, plum and cherry dessert. They drink coffee or tea with their meals and take vitamin pills.

Another oddity to me is their use of the drug eleutherococcus, a plant extract of the ginseng family. It is a stimulant, and the Soviets say it has been taken by deep-sea divers, mountain rescuers, and others to resist stress while working hard under inhospitable conditions. It is also prescribed in the Soviet Union for anemia, depression, and tuberculosis. Apparently some crews took it every morning in an attempt to increase their long-term stamina. We have no counterpart to eleutherococcus.

But to me, oddest by far is the part women play, or don't play, in the Soviet program. The Soviets got off to an early start with Valentina Tereshkova, a twenty-six-year-old parachutist who flew *Vostok 6* in 1963. Yet at this writing, more than a quarter of a century later, only one other Soviet woman (Svetlana Savitskaya) has flown in space, and she twice. This is a very strange record: of eighty-four individuals (including sixty-eight Soviet and fourteen foreign men), only two women. From 1964 to 1982 the Soviet Union launched only men.

I have asked my cosmonaut friends about this, but they have not been forthcoming. They respond, "Women? *Nyet!*" When pressed as to why *nyet,* they smile, shrug, and give me a look that says, "Collins, you bastard, you know very well why not, you just want to trap me into saying something I shouldn't." I honestly don't know why not, but apparently something went so wrong during Tereshkova's flight that the Soviets were loath to repeat it. Savitskaya must have turned out all right, or they would not have flown her twice. But even that is debatable. Cosmonaut Georgi Grechkohas intimated that Savitskaya got into trouble during a space walk and had to be rescued. Women's future in space? "No more! We love them on Earth, but in space they're no good," he told an American astronaut. The situation is a mystery to me. Their literature is contradictory. Cosmonaut training chief Georgi Beregovoi has said, "We have noticed that in training and study the whole work atmosphere and the mood in a crew of men and women are better than in a men-only one. Somehow the women elevate relationships in a small team, and this helps to stimulate its capacity for work." He certainly sounds encouraging, so why don't they use women more if they are so terrific? Beregovoi has also said, "Women are more emotional and are upset easier." General Vladimir Shatalov, the chief cosmonaut, has said he does not want to subject women to the physical strain of longer flights. Why not? The United States got off to a much slower start than the Soviet Union in selecting female astronauts because of NASA's early insistence that astronauts be experienced test pilots. However, once over that hurdle we have flown eight women as shuttle crew members, and they are well accepted by their male counterparts.

Cosmonaut Lebedev described Savitskaya's visit to *Salyut* 7. Like everyone else, she arrived in a Soyuz, but "she spent a long time in the transport vehicle getting herself ready . . . like any woman she was preening herself." Maybe Lebedev's words lost a bit in translation, but meanwhile he should stay away from Houston, where a cadre of female astronauts does everything except preen. A Soviet medical doctor, a specialist in biosatellites who had also wintered-over in Antarctica with an all-male group, told me in no uncertain terms that women should not be included on a Mars crew. Education, culture, music, chess, and conversation: all these things he favored, but women, *nyet*.

Savitskaya bristles when her role is described as bringing a pleasant atmosphere to the space station. "We do not go into space to improve the mood of the crew. Women go into space because they measure up to the job." She has no doubts about that: "At a minimum, women are equal to men in space. Women are actually better at some space tasks than men. They are better at dealing with precision tasks. They are more meticulous. They are more flexible at switching from one task to another. Men, of course, are better where heavy exertion is required." Savitskaya was commander of an all-woman crew formed in 1984 but disbanded in 1985. In addition to her space accomplishments, she is an experienced test pilot and a former world-champion aerobatic pilot.

Savitskaya has said it would be a good idea to include husbands and wives on long-mission crews. The Soviets supposedly have a group of ten female cosmonauts in training, but so far their use of women in space has in truth been closer to my friends' *nyets* than to the official protestations of support. It's a mystery that needs to be cleared up. But in the meantime I'll definitely pick women for my Mars crew.

The real evidence of Soviet success or failure in long-duration flight lies within the bodies of the cosmonauts. What do their X rays show? Have their skeletons been irreversibly weakened by a year's worth of weightlessness? Are there permanent impairments to their cardiovascular or other systems? To be absolutely certain, I suppose we must wait twenty or thirty years and see what ailments overtake these men in later life. But in the meantime the Soviet physicians project a guarded optimism. "No permanent damage" seems to be the verdict, provided the cosmonauts adhere closely to their program of exercise and diet. The bone loss seems to level off after six months, and although some irreversible damage may occur, it does not seem to be serious. The Soviets say they plan to continue long-duration flights, stretching them to eighteen months and then two years. They also want to try three or four different crews at the two-year level to make sure they understand the statistical variations among individuals.

On the day of their return to Earth after 366 days aboard *Mir,* Gherman Titov and Musa Manarov were able to walk to a helicopter with assistance, and the next day they were walking on their own. On Mars it would have been easier for them. To me this evidence indicates that a Mars voyage is feasible without the added complication of creating artificial gravity by rotating the spacecraft. But the medical community on both sides of the ocean is very cautious, hence the Soviet plans to verify that Titov and Manarov were not unusual in their response to weightlessness.

Among American space doctors it is almost as if the Soviets had never ventured into space. The data are not presented in a form they can readily understand, they complain, but most of all it is not *their* data. They were not in control, and they are reluctant to accept foreign results that they have no way of verifying. This "not-invented-here" syndrome is nowhere more apparent than in a June 1988 report by the NASA Life Sciences Strategic Planning Study Committee, which summarizes the situation as follows:

> The recent return of Soviet cosmonaut Yuri Romanenko to Earth after 326 days in space has excited great interest, as evidenced by reports in the world press. His return suggests that humans can exist for considerable periods in space and successfully readapt to conditions on Earth.
>
> Caution must be exercised, however, in drawing conclusions from a single case, particularly when the subject was unusually experienced in space missions and had been selected according to particular physiological and psychological attributes.
>
> Furthermore, the assertion that regular exercise played a role in preserving his well-being has yet to be proved. It should be noted, in addition, that his exercise program consumed four hours each day.
>
> Thus, while Romanenko's experience is encouraging, it only makes more imperative that we pursue as soon as possible the necessary studies in space to define better the physiological changes over time so that countermeasures can be rationally devised.

In other words, we have to invent an American wheel, not ride on Soviet tires. The longest NASA flight is *Skylab*'s 84 days. Our medical community wants to pick up from there, and NASA doctors say they need three more flights to prove that astronauts can safely stay aboard space station *Freedom* for 180 days. It is as if Romanenko, Titov, and Manarov do not exist, or as if the Soviets have devised some nefarious new way of calculating time.

CHAPTER

XV

PARTNERS

IN ADDITION to the Soviet Union and the United States, other countries must be counted as possible contributors to a Mars expedition early in the twenty-first century. Portugal and England are no longer in the first rank of exploring nations, but France, West Germany, and Italy have been active supporters of various space ventures. In 1973 they and seven other nations created a consortium, the European Space Agency, whose main contribution so far has been development of the Ariane rocket, a great commercial success. The Canadian government recently established the Canadian Space Agency to consolidate existing space activities formerly spread throughout the government. The Canadians built part of the U.S. shuttles: the manipulator arm that is used to grapple cargo. The Chinese have a successful rocket, the Long March, that has been approved by the United States as a launcher of American satellites—to the dismay of our own rocket industry. The Japanese have orbited scientific payloads for years and recently launched a rocket to the Moon. They say they will enter the commercial launch market in 1992—a frightening prospect for those who have competed with them in terrestrial endeavors. India and Brazil also have growing space aspirations and capabilities.

Certainly all these countries are interested in Mars, but the problem is how to include them in a rational division of labor and costs. I well remember how difficult it sometimes was on Project Apollo to get American companies to work together harmoniously. For example, the early launches of the Saturn V Moon rocket revealed a potentially dangerous vibration. The Saturn V was a three-stage vehicle, with a different manufacturer for each stage. When the entire vehicle shook and rattled, each company pointed its finger at one of the other two, and it was difficult for NASA to sort out the problem and assign responsibility for fixing it. A multinational Mars project would be an order of magnitude more complex than engineering the Saturn V, with differences not only of manufacturers but of distance, language, culture—even different systems of weights and measures.

Fortunately the space station *Freedom* can serve as a model for international cooperation. The European Space Agency has agreed to produce a laboratory module called *Columbus* that will plug into *Freedom* and that could also operate independently. The Japanese will supply a laboratory that will be a permanent part of *Freedom.* The Canadians will provide a mobile servicing center, including a manipulator arm. Within the United States, the work is being divided among Boeing, McDonnell Douglas, Rockwell International, and General Electric. NASA will be the overall manager. Before this team could get down to work, however, each member had to be assured that its own interests would be protected. On the international scene, technical difficulties aside, it took diplomatic teams three years simply to negotiate the necessary agreements. NASA will have overall management control of the station, but the work done in its laboratories will be decided upon by multinational consensus. For example, each laboratory provider can veto the use of its facility by another nation for military purposes.

A Mars undertaking will not have a military component to complicate matters, but it will be difficult enough without one. *Freedom* is fundamentally an American project, albeit with help from its allies, and so no one questions that NASA should be its overall manager. If the Soviet Union is to be included in a Mars venture, however, neither it nor the United States will be content to play second fiddle. An equal partnership is probably the solution, although I suppose there are other options, such as creating a new multinational entity to organize a worldwide effort. Certainly the United Nations could not do it, since it's organized as a debating society rather than a vehicle for getting things done. There are other organizations already in existence that might be considered. One is the Inter Agency Consultative Group, composed of the space agencies of the United States, the Soviet Union, Japan, and the European Space Agency. Another is the Space Agency Forum for the International Space Year, which includes twenty-four space agencies from around the world. The difficulty is that none of these organizations is designed for knocking heads together and getting decisions made.

Perhaps the European model of organization for developing the Ariane rocket can be used. The European Space Agency has created a corporation called Arianspace to manufacture and market a series of expendable rockets. The Ariane 1 of 1980 vintage has grown into the highly successful Ariane 4, and a more powerful Ariane 5 is planned. Arianspace is a multinational arrangement that works.

Another possibility is to use the joint Soviet-American flight of 1975 as a model. In this demonstration of détente, an Apollo and a Soyuz met in Earth orbit and the crews exchanged handshakes, food, and token gifts. Of course Apollo-Soyuz was child's play compared with the complexities of a Mars mission, but an underlying principle that ruled the first might apply to the second: make each side responsible for its own independently operated hardware, and define precisely how the pieces are to be joined together. Redundancy—and hence safety—can be provided by two Mars round-trip vehicles, one Soviet and one American. However, I don't believe that redundancy can reasonably be provided for the Martian descent, landing, and return to orbit. The weight penalty is simply too severe to provide two of each special-purpose vehicle needed to operate on or near the surface of Mars. Another possible flaw in the Apollo-Soyuz model is that it does not integrate the contributions of other nations.

No, a true international venture of this cost and complexity has never been done, and the difficulty of organizing it is certainly an argument against the idea of a multinational effort. The relationship between the United States and the Soviet Union is fragile, and historically has not been one to instill confidence that the two superpowers can agree on anything for very long, much less maintain the kind of consistent cooperation that a Mars expedition would require. To this Roald Sagdeev replies that having a joint Mars project would increase stability, giving the leaders on each side an added incentive for preserving friendly relations. Carl Sagan carries it a step further, suggesting that money to finance the Mars project be taken from the arms budgets of the two superpowers, a step that would in itself relieve tensions and thereby increase stability and cooperation.

One way to preserve the Apollo-Soyuz tradition and expand it to a multinational scale would be to assign other countries individual modules or components for which they would be responsible, but to specify that these modules would be attached to either the Soviet or American round-trip vehicle. In this arrangement, the two superpowers would act as prime contractors, each with its own stable of subcontractors. If present trends continue, the Japanese will be able to bankroll the entire project, but as Americans have already found out, the Japanese are tough negotiators, and they can be expected to demand recognition and responsibilities commensurate with the money they contribute. The European Space Agency is apt to be shorter on cash but has an enormous reservoir of experience that should be fully used. In addition to the round-trip habitat modules, the big items are the Mars descent and ascent stages, the surface habitat with a nuclear power source, and a rover for exploring the surface. The responsibility for these could be divided between Japan and Europe, with smaller contributions of components from countries such as Canada, China, Brazil, and India. Every part of a Mars mission will be vital to its successful completion, so that no matter how small a nation's contribution, it could legitimately say that without it the undertaking would not have been possible.

Of course if either superpower decides to go it alone, organizational matters become much simpler. Again, space station *Freedom* is a good start for the Americans, with NASA directing the efforts of European, Japanese, and Canadian partners. Much depends on the political future of Mikhail Gorbachev. If he prevails with *glasnost* and *perestroika* and keeps people like Roald Sagdeev at his elbow, then an international Mars venture is much more likely to happen. The idea of exploring the planets seems deeply rooted in Soviet tradition, but beyond that a Mars expedition would be a wonderful showcase, an opportunity to present the Soviet Union as a progressive force, a world leader devoting its talents to leading mankind on a peaceful pilgrimage to another world. In the face of the present Soviet overtures, the United States will find it difficult to mount a Mars expedition on its own, ignoring the Soviets. But if Gorbachev fails in his reforms, his replacement may well retreat to a more traditional Soviet worldview, one dominated by suspicion and competition. Under these circumstances a joint Mars venture would be hard to initiate and harder to sustain. But in the meantime, enthusiasm abounds on the Soviet side. For example, my friend cosmonaut Vladimir Dzhanibekov says it would be easy to put together a venture with us, each side preparing national modules that would fit together into an overall plan. Certainly when it comes to long-duration space flight, we have more to learn from the Soviets than they from us.

CHAPTER

XVI

FREE ADVICE

EXACTLY WHAT are NASA's plans for the future, with or without the Soviet Union? Much has been written about the agency, particularly in the aftermath of the *Challenger* accident, suggesting a rudderless ship, buffeted one moment by budgetary uncertainties and becalmed the next by presidential indifference. Ever since Project Apollo ended, NASA has not had a unifying goal but instead a multiplicity of programs, some of which compete with each other.

President John F. Kennedy defined NASA's mission for the 1960s in unmistakably clear terms: to "land a man on the Moon and return him safely to Earth" before the end of the decade. The man given that job, NASA's able administrator James E. Webb, worried about what his agency's role would be after a lunar landing, and in 1969 Vice President

Spiro Agnew chaired a Space Task Group that offered a choice of three long-range objectives: (1) a full-blown program consisting of a manned Mars expedition, a space station orbiting the Moon, and another station in Earth orbit that would be serviced by a shuttle vehicle; (2) an intermediate program, costing less but including a Mars mission; (3) an Earth-orbiting station and a vehicle to shuttle back and forth between the station and Earth. Agnew liked the idea of Mars, but with the Vietnam War raging neither he nor anyone else could sell the idea to Congress or the public. President Nixon picked option 3, but only part of it: he approved the shuttle but delayed the space station. That's pretty much where we are twenty years later. We have a shuttle and are waiting for a space station.

In 1985 a National Commission on Space was appointed by President Reagan and charged by Congress to formulate a bold agenda to carry America's civilian space enterprise into the twenty-first century. Chaired by Thomas O. Paine, the NASA administrator at the time of the lunar landings, the commission held public forums across the country and listened to hundreds of witnesses, expert and otherwise. It was an interesting group of fifteen, including Neil Armstrong, Chuck Yeager, and Nobel laureate physicist Luis Alvarez. The group took a long view, fifty years into the future, and their two-hundred-page report, "Pioneering the Space Frontier," is an all-encompassing list of possibilities over that period. Their findings are grouped into three broad categories: (1) "Advancing our understanding of our planet, our solar system, and the universe"; (2) "Exploring, prospecting, and settling the Solar System"; and (3) "Stimulating space enterprise for the direct benefit of people on Earth." To accomplish these goals, the nation must develop two supporting strategies: "advancing technology across a broad spectrum" and creating systems "to provide low-cost access to the space frontier."

The commission endorsed an outpost on Mars by 2015, with the following twelve milestones along the way:

- Initial operation of a permanent space station;
- Initial operation of dramatically lower-cost transport vehicles to and from low Earth orbit for cargo and passengers;
- Addition of modular transfer vehicles capable of moving cargoes and people from low Earth orbit to any destination in the inner Solar System;
- A spaceport in low Earth orbit;
- Operation of an initial lunar outpost, and pilot production of rocket propellant;
- Initial operation of a nuclear-electric vehicle for high-energy missions to the outer planets;
- First shipment of shielding mass from the Moon;
- Deployment of a spaceport in lunar orbit to support expanding human operations on the Moon;
- Initial operation of an Earth-Mars transportation system for robotic precursor missions to Mars;
- First flight of a cycling spaceship to open continuing passenger transport between Earth orbit and Mars orbit;
- Human exploration and prospecting from astronaut outposts on Phobos, Deimos, and Mars; and
- Start-up of the first Martian resource development base to provide oxygen, water, food, construction materials, and rocket propellants.

Thus the commission embraced a circuitous route to Mars, establishing first the capability to operate anywhere between Earth orbit and the Red Planet. Not only does the construction of a lunar base come first, but also the establishment of a flotilla of cycling spacecraft, repeatedly shuttling between Earth and Mars. "The people chosen for a mission will ride to an Earth-orbiting space station. From the space station they will move to a . . . spaceport and then into a transfer vehicle capable of matching orbits with a cycling spaceship. . . . On the 'cycler' they will experience a rotational gravity somewhere between that of Earth and Mars. . . . Nearing Mars after a half-year voyage, the expedition will move into another transfer vehicle to make the transition to an orbiting spaceport or a base camp. There . . . the expedition will transfer to a lander. The short trip to the surface will have as its destination either a major base camp or an outpost with a cache of supplies and equipment."

The commission recommends that before actually landing on Mars, Phobos and Deimos should be investigated, and notes that Phobos circles Mars at a distance of only 6,000 miles and is "so close that it can become a natural space station, a potential location for an early base camp. Its color is very dark, suggesting that it may be a captured asteroid rich in carbon. Similar meteoritic material indicates that nitrogen and hydrogen are found with the carbon. If that is true in the case of Phobos, it could become an ideal refueling depot for descents to the planetary surface and for the return of spacecraft to the Earth-Moon system." The report explains that Phobos and Deimos are so small they have practically no gravity; one may dock with them rather than land on them. Visiting either one would require much less fuel than fighting the strong gravitational field of Mars all the way down to its surface and back up again.

I think it certainly is possible someday to have fuel refineries on Phobos, a chain of spacecraft in an endless loop between Earth and Mars, and a series of way stations, but I don't think it is necessary to have all these things in place prior to landing on Mars. I prefer a more direct approach, with an initial landing and evaluation of conditions there before making every last decision on the infrastructure required to sustain a Martian colony. I also prefer to go to Mars without the intermediate step of colonizing the Moon, for reasons I will discuss later.

Paine's vision of the future would cost a lot of money, approximately $25 billion per year by the turn of the century. Yet as the commission points out, "the percentage of the U.S. Gross National Product invested on the space frontier would remain below half of the peak percentages spent on the civilian space program during the peak Apollo years. In view of the increasing significance of space and the critical economic importance of U.S. scientific and technological preeminence in the next century, we believe that these estimated levels of expenditure are reasonable in relation to the expected benefits to our nation and the world." Far from draining our national coffers, the Apollo years were a golden age of technology, with advances across a wide spectrum that benefited American industry and kept us competitive on the world market. Less than one-half of Apollo-level expenditures certainly makes sense to me if we are to survive as leaders in a high-tech, increasingly competitive world.

"Pioneering the Space Frontier" makes for fascinating reading, an unusual attribute for a report written by a commission. But despite my affection for it, I don't think it is a very useful document for NASA to use as a planning guide. It is too comprehensive and covers too many things over too long a period. There are too many ways for a bureaucrat to follow the Paine report without getting much done. "Less is more," as the architects say. But at least Mars is the centerpiece of the report; even if it is half buried by other objectives. Unfortunately the report was issued in May 1986, in the wake of the *Challenger* explosion, and it never got the attention it deserved.

NASA liked the Paine report because it painted a very rosy picture of that agency's future, but at the same time Dr. James Fletcher, NASA administrator at the time, recognized that he needed something a little more tangible, and more immediate, to guide his agency through the remainder of the twentieth century. At that time NASA was grappling with words like "vision" and "leadership initiatives." Fletcher brought Sally Ride, America's first woman astronaut, to Washington in 1986 to organize a study on NASA's future in the aftermath of the *Challenger* disaster. On her desk was not only the Paine report but also the Rogers commission investigation of the *Challenger* accident. The Rogers group had put together a long list of corrective actions required of NASA before the shuttle could fly again.

At about this same time I was a member of the NASA Advisory Council, a group that meets several times a year, listens to NASA's leaders, and gives them free advice. I was put in charge of a task force to look into NASA's goals. We were also asked to write a few words on that elusive quality called leadership. Our report was intended for NASA's internal use, with an eye toward any assistance it might give Sally Ride in her larger study.

Ride had a tough job, she wrote, a balancing act acknowledging

> the visions of the National Commission on Space but faced with the realities set forth by the Rogers commission. NASA must respond . . . to both while recognizing . . . reasonable fiscal limits. Two fundamental, potentially inconsistent views have emerged. Many people believe that NASA should adopt a major, visionary goal. They argue that this would galvanize support, focus NASA programs, and generate excitement. Many others believe that NASA is already overcommitted in the 1990s; they argue that the space agency will be struggling to operate the space shuttle and build the space station, and could not handle another major program.

In August 1987 Ride issued her report. It selected four tidbits from the Paine smorgasbord and offered them to NASA as the next steps to be undertaken: (1) Mission to Planet Earth; (2) Exploration of the Solar System; (3) Outpost on the Moon; and (4) Humans to Mars.

"Mission to Planet Earth," her report stated, is "a program that would use the perspective afforded from space to study and characterize our home planet on a global scale." Having the ability to orbit the Earth once every ninety minutes is a powerful tool that has not yet been fully developed. I live close to the ocean, and therefore it is not surprising that one of my main concerns is the effect of the pollutants we pour into the sea. We have always considered the sea's capacity infinite, but now we find that our Styrofoam cups end up circling endlessly in the Sargasso Sea, that parts of the Mediterranean are little better than a cesspool: it is clear there are limits to the punishment our oceans can absorb. The first step in fixing this problem is to assess the damage by pinpointing the sources of pollution and tracing its dispersion offshore. Dr. Gifford Ewing, an oceanographer, has noted that our conventional method of taking measurements, by instruments dropped from boats, is like reading a newspaper by placing it flat on the floor, sticking pins in it, and then decoding the ink stains on the pins. Sensors aboard a satellite would at least allow the front page to be read at a glance, although admittedly space detectors do not have the ability to penetrate the depths of the oceans. Similar sensors have the ability to measure certain conditions in our atmosphere and on the fragile surface of our planet: this is what Mission to Planet Earth is all about. The Ride report proposes a network of nine orbiting platforms.

Exploration of the Solar System, as described in the report, breaks down into three robotic missions. The first is called CRAF—Comet Rendezvous Asteroid Flyby. The comet selected is Tempel 2, and the asteroid is Hestia. After a 1993 launch and a six-month cruise, the CRAF vehicle's visual and infrared sensors would examine Hestia's surface composition and structure, and then continue on to Tempel 2, where it would shoot two penetrators into the nucleus of the comet for detailed measurements. Scientists hope that CRAF would reveal fundamental information about the beginnings of the Solar System.

The second mission in Exploration of the Solar System would be a visit by *Cassini* to Saturn and its largest moon, Titan. To me Titan is the most fascinating moon in our

Solar System. Its atmosphere, although hardly Earth-like—it is thicker than Earth's and made mostly of methane—does contain nitrogen and other gases that were present in our own atmosphere billions of years ago. Its surface is very cold, perhaps 300 degrees below zero, and is an inhospitable mixture of ice, rock, and metals. But it also includes a dusting of various organic molecules, including hydrocarbons and single carbon-nitrogen compounds. Titan has a hot core. Between the icy surface and the hot interior, who knows what kind of subterranean lakes might exist, and what kind of life might have evolved in them? Only recently have we discovered, deep in our own oceans, strange life forms nurtured in water warmed by volcanic vents. These creatures, including tube worms six feet long, exist by using a hydrogen sulfide metabolism independent of sunlight. Launched in 1998, *Cassini* would probe the atmosphere and surface of Titan and take scientific measurements of Saturn itself.

The third part of Exploration of the Solar System would be a Mars rover/sample-return mission. It would involve a soft landing on the Martian surface, deployment of a computerized, "smart" rover to collect rock and soil samples, and delivery of these samples back to Earth orbit for analysis on board the space station. Three such voyages would be included, two launched in 1996 and one in 1998.

Sally Ride's third selection, Outpost on the Moon, is divided into three parts: Phase I: Search for a Site (1990s); Phase II: Return to the Moon (2000–2005); and Phase III: At Home on the Moon (2005–2010). Phase I would map the surface and search for water at the poles. Phase II would put people on the surface for a week or two, setting up a pilot plant for extracting oxygen from lunar rocks. Phase

III would be a base with closed-loop life-support systems. Up to thirty people would work there for months at a time. I find returning to the Moon the least appealing of the Ride report's four selections. I remember the Moon from a distance of sixty miles seeming a sterile, lifeless rock pile, a monotonous place much less appealing than our sister planet Mars. Perhaps, because I personally have no desire to go back there, my thinking has become warped, but to me returning to the Moon is an expensive and unnecessary detour on our way to a much more interesting place.

Sally Ride's fourth option—Humans to Mars—is my favorite. Its first phase would consist of robotic missions to gather rock and soil samples, in addition to performing geochemical analyses and mapping the surface. This information would permit selection of a landing site. The second phase would "establish an aggressive space station life sciences research program to validate the feasibility of long-duration spaceflight. . . . An important result would be the determination of whether . . . vehicles must provide artificial gravity." The third phase would consist of landings, three of them, "fast piloted round-trip missions. . . . These flights would enable the commitment, by 2010, to an outpost on Mars."

The flights that Sally proposes are called sprint missions, and they divide the hardware in a way that makes me nervous. The idea is to send supplies on ahead, on a slow, fuel-efficient trajectory, and then have the crew sprint after the supply vehicle, catching it in Martian orbit. The advantage is that by making the crew's spacecraft smaller, it could go faster (round-trip time estimated at one year, with ten to twenty days on the surface). The disadvantage is that if the rendezvous were botched somehow, the crew would be dead.

Granted they would not leave Earth until their supply vehicle was ensconced in orbit around Mars, but still . . . I think there has to be a better way. And once again it is interesting to note how little attention NASA pays to the Soviets. At the time the Ride report was being written, the Soviets had kept one three-man crew up for 237 days and Yuri Romanenko was already airborne on what would be his 326-day flight. Yet NASA still insists, as in the second phase of Humans to Mars, that it must establish a research program aboard *Freedom* to "validate the feasibility of long-term space flight."

The Ride report summarizes the Humans to Mars option as follows: "A successful Mars initiative would recapture the high ground of world space leadership and would provide an exciting focus for creativity, motivation, and pride of the American people. The challenge is compelling, and it is enormous." Nowhere in the Ride report is there mention of the Soviets. Sally's instructions from NASA administrator Jim Fletcher were to leave them out, to limit the scope of the study and to make it less complicated. Including the Soviets would certainly have caused Ride to put a different twist on the summary.

The report of my own group, the Space Goals Task Force of the NASA Advisory Council, essentially said, "Mars, but . . ." We thought that the human exploration of Mars should be our primary goal, and so stated, but we recognized that a number of preliminary steps were required, and that NASA could not abandon its other responsibilities in an all-out quest for Mars.

We began by noting that the National Space Policy of 1982 includes a mandate to "maintain U.S. space leadership." We thought that our leadership should center on human exploration, space technology, and space science—and required a proper balance among the three. We did not think that leadership needed to be achieved in every element of the three areas, but certainly in the key ones. Our group was concerned that during the past decade our technology base had eroded; we felt we needed to restore our industrial capacity to master the newest techniques for building space machines efficiently and reliably.

As preliminary steps, we mentioned returning the shuttle (which was then grounded) to safe-flight status and developing an expendable launch vehicle to complement the shuttle. These two launchers would allow the United States to pursue a variety of scientific missions, including robotic forays to Mars. We endorsed the space station. We put in a plug for commercial ventures in space.

But Mars was the centerpiece. Mars, we wrote,

> for centuries an object of intrigue, stands out as the one entity most likely to capture widespread enthusiasm and support, while pulling considerable scientific and technical capability in its wake. . . . We should make exploring and prospecting on Mars our primary goal, and should so state publicly. A bold goal, clearly stated, will result not only in increased public awareness of NASA, but will serve the internal purpose of providing a stimulus to focus and clarify programs, and to enhance productivity and efficiency.

We liked the word "prospecting" because it implied a permanent, or at least a long-term, presence on Mars, including activities that might prove beneficial from a scientific or commercial point of view. We did not rule out using the Moon as a stepping-stone to Mars, but neither did we endorse the idea. We thought it should be studied.

In regard to the international aspect, we said, "Extending the human presence to Mars should be a peaceful enterprise done in the name of all humankind. Under appropriate conditions, not only the Soviet Union, but other qualified nations should be invited to participate with the United States and to pool scarce resources. As President Reagan has said in regard to human exploration: 'We hope our friends and allies will join us in this great adventure!' "

One of Sally Ride's recommendations was that NASA establish an Office of Exploration, and that has been done. Its first major report, "Beyond Earth's Boundaries," was issued as its 1988 Annual Report to the Administrator of NASA. It outlined four case studies: (1) a human expedition to Phobos; (2) a human expedition to Mars; (3) a lunar observatory; and (4) a lunar outpost leading to Mars.

In summary, by 1988 NASA had been presented with at least four views of its future, ranging from the four pages put together by my group all the way to the voluminous report by the National Commission on Space. Mars features prominently in all of them, and the Moon in three. But to all, and to any outsider who asks, NASA steadfastly replies that it needs space station *Freedom*. If a space station seems a strange response to Moon, Mars, and science in general, then perhaps a review of NASA's history will put it in better perspective.

NASA has always been a "big-ticket" agency, with one dominating project, one that keeps most of its work force gainfully employed. First was Mercury, to put John Glenn and cohorts into orbit. Then Gemini, to prove that going to the Moon was feasible. Then Project Apollo itself, followed by its offshoot, *Skylab*. The shuttle then took over and has dominated since. Now the shuttle impetus is disappearing, and NASA needs another centerpiece. For a number of logical reasons, it has picked the space station. Estimated to employ fifty-two thousand people in fourteen states, *Freedom* will carry NASA for the next half-dozen years at least. It will prevent disintegration of the agency by keeping a stable work force, employing elements from all NASA's centers in the design, manufacturing, and operation of a permanently manned station. It will bring bureaucratic stability, if not excitement. And *Freedom* can legitimately be considered a first step in the direction of a Moon colony or an expedition to Mars.

However, not looking beyond the space station has also had its drawbacks. *Freedom* will be a multipurpose facility, and therefore a compromise. To some extent that broadens its constituency and makes it more appealing, but unfortunately the space community is divided into parochial groups. Some who intend to use the station as a laboratory want its design optimized for their particular experiments. Others want their own platforms in the sky and see *Freedom* as a competitor that will siphon off NASA's funds and attention, leaving them stuck on the ground. For example, proponents of Mission to Planet Earth point out that *Freedom* will be in the wrong orbit for their purposes, crisscrossing the equator instead of flying over polar regions. Those who want to melt and cool crystals in perfect weightlessness don't like the idea of a human crew banging around disturbing things, or operating vibrating equipment like a centrifuge. If a politician wants to save some money he or she can produce an impressive array of space experts who will testify that *Freedom* isn't what they had in mind. The next step in this process is to accuse NASA of confusion and disarray and to cut its funding. That's what has happened to the space station so far. Without a unifying goal beyond the station, and without a compelling *raison d'être, Freedom* has become an albatross around NASA's neck rather than a talisman of future success.

As the next two chapters will show, President Bush has taken some steps to depict *Freedom* as a necessary preliminary to returning to the Moon and landing on Mars.

CHAPTER

XVII

WHY?

If Mars is to become NASA's long-term goal, the reasons must be clearly understood and laid out in much greater detail than was the case for the Moon argued by John F. Kennedy. Four months into his presidency, on May 25, 1961, Kennedy told a Joint Session of Congress: "I believe that this nation should commit itself to achieving the goal, before this decade is out, of landing a man on the Moon and returning him safely to Earth." He included a few reasons: ". . . impressive to mankind . . . important for the long-range exploration of space . . . time for this nation to take a clearly leading role in space achievement, which in many ways holds the key to our future on Earth." But his audience understood, and he did not have to spell out, his real reasons and the immediacy of his concern. Five weeks earlier, two events had badly shaken the world's confidence in the United States: a CIA-supported invasion of Castro's Cuba had failed on the beaches at the Bay of Pigs, and a man named Yuri Gagarin had become the first to orbit the planet. As the *Washington Post* noted, "Ever since the Cuban invasion fiasco the bloom has been off the bright rose of the early days of the new Administration." Kennedy needed an achievement to offset these failures, and his experts had told him NASA could beat the Soviets to the Moon. The birth of Project Apollo was as simple as that.

Of course, in a technical sense Apollo was far from simple, but complicated as the engineering was, at least NASA was allowed to proceed with a single-minded dedication, without having to justify its existence every step of the way. With Kennedy's voice ringing in its ears, and the decade passing quickly, NASA wasted no time. A clear goal, an immutable deadline, and enough money: those were the key ingredients of Apollo's success. Apollo was perceived as a convincing demonstration that the United States could do what it set out to do, under some circumstances at least, and the success of *Apollo 11* was greeted around the world with great jubilation.

I was privileged, in its immediate aftermath, to visit more than twenty foreign countries, and the reaction in all was the same (and surprising to me). No one said, "Well, you Americans did it." Adversaries and friends responded in the same way, with admiration and respect. They all said, "*We* did it!" There was a genuine outpouring of pride in a *human,* rather than a nationalistic or technological, accomplishment: the first extraterrestrial expedition, without weapons, and for the benefit of all mankind.

Having said all that, however, it is also true that the American public quickly tired of subsequent lunar landings. It was like the same Super Bowl being broadcast time after time. Also, it was not clear what the average citizen gained from Apollo, although econometric studies indicated that a dollar spent there resulted in between five and seven dollars' worth of benefits permeating other parts of the U.S. economy. But when Apollo was over, it was over. It was not perceived as a gateway to the future but as an end in itself. A Mars mission, if constructed along Apollo lines, will also be so considered.

At the time of Apollo the United States was the world's greatest creditor nation. Today, to our shame, we are the largest debtor. A by-product of our gigantic federal deficit is the Gramm-Rudman-Hollings Deficit Reduction Act that arbitrarily cuts spending if certain targets are not met. In such a vineyard does George Bush labor, required to do as much pruning as planting. The nineties will be much different from the sixties. However, before a new generation thinks that its problems are unique, it might read President Kennedy's 1961 State of the Union message. "The present state of our economy is disturbing . . . seven years of diminished economic growth . . . business bankruptcies have reached their highest level since the Great Depression . . . farm income has been squeezed down by 25 percent . . . 5½ million Americans who are without jobs . . . our cities are being engulfed in squalor . . . 25 million Americans living in substandard homes . . . we lack the scientists, the engineers, and the teachers our world obligations require." Four months after acknowledging these conditions, somewhat similar to today's, Kennedy proposed Project Apollo.

To me, the *why* of a Mars mission is rooted in the history of our planet and of this nation. I don't know whether exploration is in our genes, although I certainly think curiosity is, and the two are closely linked. At any rate, whether it is an inherited or acquired characteristic, most people have always gone wherever they could go. Some tribes have not, being content to live out their lives sealed off in remote valleys or hidden in rain forests, but even most primitive civilizations—Polynesians in their canoes, nomads on their camels, bushmen on foot across the Kalahari—have generally been wanderers. Certainly our European forebears, especially the Portuguese, Spanish, Italians, and British, were obsessed with reaching the farthest corners, if a globe can be considered to have corners. Many of their voyages were similar in duration to a Mars trip, and early explorers such as Ferdinand Magellan endured hardships that I hope will never be replicated in space. The great explorers were willing to put up with such long-term privation for a variety of reasons. Patriotism, ego, religion, inquisitiveness, greed—all played a part. When asked what he hoped to find in India, Vasco da Gama replied, "Christians and spices." There will be no Christians waiting on Mars and probably no spices, but eliminating religion and commercial return (at least for the first few landings) is not necessarily bad. The urge to go, to see, to touch, to smell, to learn—that is the essence of it, not to mention the exhilarating possibility of encountering something totally unexpected.

The ocean voyages of the littoral European nations eventually charted the outlines of the continents and established outposts on the eastern seaboard of North America. They were squalid settlements, places like Plymouth and Jamestown, whose citizens looked toward the interior. As soon as they could they pressed westward over the Appalachians and out onto the great interior plains. Then to the Rockies, and finally to the Pacific. Conditions were miserable in the early settlements, with poverty and disease rampant, and life on the trail was no better. Some chose to stay, some to travel. No one suggested that the towns first be elevated to a certain standard of hygiene or nutrition as a precondition to exploration.

The first expedition for which Congress appropriated funds was that of Lewis and Clark. In 1803 President Thomas Jefferson sent a confidential message to Congress requesting $2,500 "for the purpose of extending the external commerce of the United States by means of an expedition." Improving trade with the Indians along the route was the ostensible reason given by Jefferson, but he also had not abandoned hope of finding a northwest passage, a riparian route connecting the Missouri River with the Pacific Ocean. Furthermore, Jefferson was fascinated with the scientific harvest to be reaped. In his detailed instructions to Lewis, he discussed such matters as compass sightings, prevailing winds, mineral identification, "volcanic appearances," plants, flowers, birds, reptiles, and insects. As a consequence Lewis and Clark recorded flora and fauna not previously known in the East, and they shipped back to Jefferson large collections of dried plants, seeds, cuttings, skins, horns, and bones—and even a live prairie dog. But the main benefit of the expedition was increased geographic knowledge of the West. Even more fundamental, a precedent was established in this country for government sponsorship of scientific ventures. Unlike most science today, the Lewis and Clark expedition was shrouded with secrecy. Also, like a lot of other government enterprises, it experienced a monumental cost overrun—the final price tag was more than $28,000. But it was money well spent.

It is undeniable that Thomas Jefferson represented a special case of enlightenment—a man of science as well as of politics. But regardless of our political leadership, the spirit of the frontier, of venturing into the unknown, has pervaded the history of this nation. It is part of the American character. In a small but very important way, exploration defines us as a nation. Sometimes we have been too venturesome in asserting our presence, at least in the field of geopolitics. But in art, literature, science, and commerce, our willingness to take a chance on new ideas and new places has served us well. The expedition of Lewis and Clark is just one in a long series. In similar fashion, the expeditions to the Moon repaid the nation several times over, both directly in terms of national esteem, vigor, and enthusiasm, and through spin-offs—new products and enhanced attitudes—to other parts of our economy. If nothing else, Apollo intensified the national debate on priorities and future directions. How many times have we heard, "Well, if we can put a man on the Moon, why can't we . . . ?" Apollo is used as a standard, sometimes an inappropriate one, for what can or should be done. It is also used as an object of ridicule, of tax dollars frittered away on the Moon and its rocks, just as I am sure people complained about Lewis and Clark's bringing back a prairie dog instead of gold.

Of course none of the $24 billion that went into Apollo was spent in outer space. The bucks were spent in the United States on high-technology jobs that helped make our industries competitive in the world market. Since the

Apollo days we have suffered a steady decline in our exports. Electronic products, once an American specialty, are now flooding in from Pacific Rim countries. The aerospace industry, having surpassed agriculture, is presently the number one contributor to a favorable balance of trade (around $17 billion per year), but if aerospace goes the way of electronics, our trade-deficit problems will be greatly exacerbated. The majority of our aerospace exports consist of the big Boeing and McDonnell Douglas airliners, but aviation and space technology are closely intertwined. For example, a new strong and light composite material developed for one can be used in the other. In space there is a relentless quest for improved performance, safety, reliability, and efficiency. The stringent requirements of a Mars mission will accelerate the development of technology that will be applicable to the aviation industry and to American products in general. We tire of faulty products, things that break. It's bad enough when something conks out on the ground, and it's a very serious matter in an airliner at 35,000 feet, but it's an absolute disaster months away from Earth on a Mars expedition. If one segment of American industry is forced by the very nature of its job (Mars) to achieve the highest possible standard of excellence, then that intangible attitude and the tangible products resulting from it will permeate other parts of American industry. After twenty years Apollo spin-offs are still contributing to such unlikely fields as clothing and equipment for fire fighters. In another area of concern to all of us, NASA claims that medical devices deriving from the space program have had a total economic impact of $1.8 billion since 1973. Benefits such as these are possible because of work NASA has done in computers, microelectronics, electrical power, inertial systems, computational fluid dynamics, and thermal control.

More directly, American and foreign products circle overhead with increasing frequency as the market grows for commercial satellites used in communications, weather forecasting, and Earth surveying. As a satellite launcher the United States is in danger of being excluded from a field in which it was once paramount. The French Ariane booster has taken a great deal of business away from us (especially during the time the shuttle was grounded), and the Chinese are having some success in marketing their Long March rocket. The Soviets have offered the Proton to the world market at bargain-basement prices, and—probably most ominous—the Japanese are preparing to enter the fray after years of meticulous preparation (as usual). As a recent NASA advisory group pointed out, "A single $100 million launch contract is equivalent in economic terms to the import of 10,000 Toyotas."

In 1961 President Kennedy was goaded into action by what he perceived as a grave Soviet threat. Yuri Gagarin's flight had dramatized *Sputnik*'s earlier warning of a missile gap. Kennedy responded with a Moon race. Today the threat is more subtle, and it comes not just from the Soviets but from all around us, even from France, with whom we have rarely competed directly. That the threat is economic rather than military in no way lessens its importance to U.S. citizens. If present trends continue, the 1990s will bring some lean times for us Americans. One difficulty in preparing for them is that our schools seem to be deteriorating in terms of the math, science, and engineering students they produce. In a recent six-nation survey U.S. children scored last in math and near the bottom in science. This does not bode well for our ability to compete in the future. Somehow we must motivate students to enter these technical fields and pursue their studies with a vigor that matches that of their Asian and European counterparts. The Apollo program served as such a catalyst, but as NASA has fallen into disarray, so has interest in a space career among the young. The correlation is very high between Apollo funding and the production of science and engineering graduates. The number of doctorates in these fields produced during the 1960s and 1970s is almost exactly proportional to NASA's appropriations. A Mars project would rejuvenate NASA, and would carry in its wake a new generation of scientists and engineers with the education necessary to succeed not only in space careers but in other high-tech endeavors as well.

With our gargantuan national debt to service, federal money will be tight for years to come. Space is apt to get muscled out, because it is in that shrinking category called discretionary spending—funding not mandated by law. Under the congressional appropriations apparatus, NASA is assigned to a subcommittee in which it competes directly with the Department of Housing and Urban Development and the Veterans Administration, two agencies with large and vocal constituencies. From this arises a form of either-or logic best exemplified years ago by then-Congressman Ed Koch: "I just for the life of me can't see voting for monies to find out whether or not there is some microbe on Mars, when in fact I know there are rats in the Harlem apartments." So it is rats versus Mars, and it is hard to explain why money should be taken from one and given to the other.

Certainly the government has an obligation to its citizens in maintaining general standards of health and hygiene, but the government can't assume the responsibility for driving every rat from every apartment. Perhaps the best way to get rid of the rats is by creating an optimistic climate in this country, a can-do spirit that says properly motivated individuals and groups can achieve reasonable goals—big things and little things. "If we can land a man on the Moon . . ." Furthermore, I think a government has an obligation to its citizens not only to fight today's battles but to look to tomorrow, to assure that the country is headed in a direction that will create higher living standards. Prudent farmers save a little seed corn. Balance of trade is a key indicator of economic success, and balance of trade in this country is tied predominantly to agriculture and high technology. Government investment in space exploration is a stimulus to our high-tech economy that may in the long run remove more rats from Harlem than subsidized housing grants will. America has been synonymous with opportunity, and the ability to go out into space for knowledge or profit continues that tradition.

There is also a scientific bonanza waiting on Mars: comparative planetology, the scientists call it. They would like to study the history of the Solar System and the evolution of the planets by comparing evidence found on Earth, the Moon, and Mars. The history of Earth's climate, for example, can be partially decoded by examining ice tubes bored from the Antarctic crust. It would be extraordinarily helpful to be able to compare these with ice bored from the Martian poles. Today Earth's atmosphere seems susceptible to ozone holes, particularly a large one near the South Pole. Does Mars show similar evidence? How recent is the vulcanism on Mars? What caused it to stop? What happened to all the water that carved out deep channels? Could a similar process occur on Earth? What caused these two planets, similar in so many ways, to evolve so differently? To scientists these questions are fundamental to a better understanding of our own planet.

But even more fundamental to me than science or the pocketbook are matters of the spirit. Space exploration is a victory of technology, but only in the same narrow sense that a new thought is a victory of neurons and synapses. When we explore the Moon or Mars, we really explore ourselves and learn more accurately how we fit in. For centuries we were guided by the ideas of Ptolemy, the Egyptian astronomer who took elaborate pains to rationalize the motion of Sun, planets, and stars in terms of an Earth-centered coordinate system. Ptolemy taught that the Earth was the center of everything. Galileo set us straight and today every schoolchild has had explained that the Sun is the local center about which we and the other planets revolve. But psychologically we still cling to the Ptolemaic view, and no wonder—as long as we stay here on Earth, we are the de facto center of everything that matters. We say that the Sun goes down and comes up. It really does not; the Sun hasn't gone anywhere; it is simply that our spot on this planet has rotated into the shadows or back out into the sunlight. Do you have trouble remembering whether Europe is ahead or behind the U.S. in local time? Go to the Moon and look back and you will see it clearly: first Europe, then a few hours later North America, swings around into evening and night. You don't have to go to the Moon to understand this, but you do—or at least I did—truly to *feel* it. It may be trivial, this matter of watching the Earth from afar. But I think not. I think everyone who has flown in space, even just a hundred miles away, has come back with a profoundly different way of looking at the home planet. A social scientist named Frank White has written an entire book on the subject entitled *The Overview Effect*. In it White writes that "going into space is not about a technological achievement, but about the human spirit and our contribution to universal purpose. Space . . . is a metaphor of expansiveness, opportunity and freedom. More than a place or even an experience, it is a state of mind. It is a physical, mental, and spiritual dimension in which humanity can move beyond the current equilibrium point, begin to change, and eventually transform itself into something so extraordinary that we cannot even imagine it."

I don't know that this transformation will take place, but going as far away as Mars would certainly test the hypothesis. Reporting on his orbital flight aboard the U.S. shuttle, Prince Sultan bin Salman al-Saud said, "The first day or so we all pointed to our countries. The third or fourth day we were pointing to our continents. By the fifth day we were aware of only one Earth." I can attest that from the Moon, some 236,000 miles away, I was drawn to home like a magnet. But home was not my home in Houston, where my wife and children were; it was the entire blue and white marble. I didn't care whether Texas or Australia was pointed at me; I couldn't take my eyes off it. It was exquisitely beautiful, bright and shiny in the sunlight. I could see no political borders, or even signs of life. It seemed pristine, which it is not, and fragile, which it is. The notion of fragility crept quietly into my consciousness; at first it was all beauty, then gradually "beautiful but fragile" took over, and today I think I have an intensity of concern for this planet, and especially its oceans, that I could not have generated before. I feel strongly that more people, especially political leaders, need this vantage point. I know it sounds conceited, but I am a better citizen and person for having flown in space. It is a humbling experience to see the Earth from afar, and I feel extremely fortunate to have enjoyed so privileged a view. I wish I could do a better job of repaying the planet. At least I can pick up a few beer cans on the beach.

From Mars, who knows? The reaction of humans there may be totally different, because the Earth will certainly not be a visual spectacle beckoning through their windows. Rather, Earth will be just another bright blob of light in the black sky. Will our crew members crane their necks, seeking it out anxiously, or will they acquire a celestial *savoir-faire* that makes them indifferent to it? Might they even, like one of Mark Twain's characters, refer to it disdainfully as the Wart? I don't know, but in any case they will see it totally differently than we surface crawlers do, and that may have some interesting ramifications. For instance, comparative planetology in a social rather than a scientific sense. One cannot take a close look at the scarred, Sun-seared peach pit that is the Moon without realizing what a delightful treasure the Earth is, a place of rose gardens and misty waterfalls. That will be true of visitors to Mars too, I suspect. But Mars can be a lot closer to a happy home than can the Moon, and travelers to Mars may become attached to some of its sights and compare them favorably to similar features on Earth.

With a Martian surface area about equal to that of all the dry land on Earth, there will be plenty from which to choose. The gigantic volcano Olympus Mons, for example, must surely be one of the most unforgettable places in the

Solar System. Likewise the extraordinary canyons of Valles Marineris, in places nearly four times deeper than Earth's Grand Canyon. The salmon pink sky, wispy clouds, the frozen polar regions: Mars has infinitely more to offer than the Moon, a place only a geologist could love.

And when it comes to actually colonizing the Red Planet, our Martian settlers may become fiercely proud of their locale, just as I have recently become of the Outer Banks of North Carolina. For the first time, the psychologists who predicted a "breakaway" phenomenon among space travelers may find themselves right. Our new Martians may not look back. They may establish ways that are different from Earth ways, they may do things better. Certainly they will have some powerful advantages. For example, the first to arrive should be a close-knit, cohesive group with no "us versus them" thinking—unless it is "us Martians" versus "them Earthlings." And with no indigenous diseases, no traditional enemies, no borders, no lack of real estate, no accumulated wealth, and a common necessity to struggle against the harsh environment, Martian settlers will not have a lot of the excess baggage that impedes friendly relations among Earthlings. They will have a fresh start, these people *of* the Earth but not *on* the Earth. All this sounds a bit utopian, I realize. On Earth, expedition members have turned violently against each other when things went wrong. Murder, rape, cannibalism—all have been recorded in small groups, along with heroic deeds. Anthropologists speak of the alpha male and the constant struggle within a group to displace him. The group purges itself, but it can be a painful and even deadly process. Humans on Mars will still have their dark side, and things won't be easy for the early settlers, but I can't help but feel optimistic about their ability to establish a colony that accepts the finest components of human behavior and rejects the basest. Selection and training of the first groups will be important in making this happen.

Mars will not pull at one's body the same way Earth does. At one-third their Earth weight, our Martians may find their habitats and their habits gradually changing. A return to Earth would certainly be a disappointment in this one respect at least. They would be light people visiting heavy people. Probably they would not like being squashed flat while making love. A visit to one of their two tiny moons would pose the opposite problem. On Phobos the gravity field would be about a thousandth as strong as on Earth, and even less on Deimos. A base there or a visiting spacecraft would have to be anchored to stay in place.

As seen from the surface of Mars, neither moon will dominate the sky. Deimos will appear simply as a bright star, while the larger Phobos will seem about one-third as large as Earth's moon. The two move in opposite directions, Phobos passing from west to east in about five hours while Deimos creeps imperceptibly toward the west, taking sixty hours to cross the sky. Although the Soviet *Phobos 2* probe casts doubt on the theory, some scientists believe that Phobos is a visitor from beyond Mars, an asteroid captured by the planet's gravity. Its color indicates that it may be composed mostly of material found in those stony meteorites called carbonaceous chondrites. It may have formed very early in the evolution of the Solar System and remained essentially unaltered since it condensed from the solar nebula, around 4.6 billion years ago. If so, Phobos is an object worthy of scientific investigation independent of Mars. Some engineers have also pointed out that it may be rich in water and organic compounds and could be used as a source of rocket fuel, eventually reducing the amount of propellant brought from Earth.

At any rate, our new Martians will have a variety of surroundings not available on the Moon, and some visual treats and physical sensations that may match the best Earth has to offer. Their high-tech colony will have to grow using un-Earthly techniques. Their rivalry with Earth, their admiration or criticism of Earth—any of these consequences of a human presence on Mars can only help the home planet, help it to understand itself better and to write a prescription for change. If it is true that someone who doctors himself has a fool for a patient, then perhaps a planetary consult might be in order, a second opinion from a very different place.

I don't mean for Mars to be an escape valve. For more than a hundred years we have debated the theory of Malthusian expansion, and more recently the Club of Rome has attempted to define some limits to growth. On Earth the supply of food may always lag behind the production of new human beings, but Mars cannot be counted on to alter this equation, either by supplying resources to Earth or by siphoning off Earth's excess population. The distance is too vast, the cost too high. But living in close quarters under a dome, squandering nothing, recycling products to the maximum, Martian colonists may very well find new antidotes to Malthusian pressure.

I had to see a second planet—the Moon—to fall in love

with the first. Perhaps we need to live on a third—Mars—to go beyond love to a successful marriage with the first. How would people on Earth react without any weapons, for example? Since the dawn of history our hunting ancestors have had them, and today, as Carl Sagan has written, "The U.S. and the U.S.S.R. have now booby-trapped the planet with nearly 60,000 nuclear weapons." There would be no need to carry weapons to lifeless Mars, and it would be easy to prevent their transport from Earth. Therefore, for a while anyway, a modern society would be totally without them. What could we Earthlings learn from this? How would Martian disputes be settled? In my opinion, escaping from hydrogen bombs is not a valid reason for leaving Earth, but learning to live without them may be.

Planets are all we have left to explore, in a physical sense, except for a bit of roaming on the sea bottom. Only the inner planets are accessible now. Mercury and Venus are impossibly hot, and we have already done the Moon (wow, will some of my compatriots argue with that!). Therefore, if we have a spiritual need for a new frontier—and I believe we do—Mars is it. If there is a migratory drive within us—and I believe there is—it will lead us to Mars. If there is an extraterrestrial imperative, Mars is surely the next logical stepping-stone on the endless journey to the stars. Our bodies are no more than star stuff that coalesced along with the Earth, debris from the original explosion that created everything we know. We won't stay here. Call it genes, character, culture, spirit, ethos: by whatever name, it is within us to look up into the night sky and be curious, within us to commit our bodies to following our eyes.

Exploration: I don't want to live without it; I don't want to live with a lid over my head. What will exploration gain us, beyond allowing us to dream? T. S. Eliot said it better than I can: "We shall not cease from exploration and the end of all our exploring will be to arrive where we started and know the place for the first time."

CHAPTER

XVIII

WASHINGTON AND SIMILAR PLACES

THE DECISION to go to Mars will be made for political reasons. The economic payoff is too vague and too long-term. Scientific research on Mars will be grand, but science alone has never been seen as justifying such an expensive project. (Texas's supercollider, at $5 billion, is the champion in that category.) We will go to Mars because some people in Washington, D.C., think it is a good idea. Who are these people, and what might convince them? They are the president and his team in the executive branch, and the leaders of Congress. As usual in our democracy they will listen to their constituents, the press, and their consciences before deciding.

The message they are getting from their constituents is a

mixed one. Polls yield wildly differing results, depending on how they are worded. On the negative side, when Americans are asked what might be cut from the budget, space usually ranks up there with foreign aid. And yet when asked directly about space, Americans generally have a very favorable reaction. In January 1988 *Government Executive* magazine published the responses to a Roper Organization poll that asked, "For the following budget items, should we spend more, the same, or less?"

	More (%)	Same (%)	Less (%)
Taking care of the homeless	68	21	6
Education	66	24	5
Social security benefits	63	27	5
Health	63	25	6
Aid to the poor	63	21	11
Job creation and training	61	25	8
Environment	47	34	11
Agricultural price supports and subsidies	40	26	21
Science and basic research	39	40	13
Improving mass transit	36	43	14
Aid to cities/states	27	40	23
Space exploration	17	34	42
The military, armaments, and defense	14	32	48
Foreign aid	5	18	71

Shortly after this poll, the Public Opinion Laboratory at Northern Illinois University took a poll that separated people keenly interested in space (space attentives) from all others and asked whether they agreed with statements about space exploration. Sample results:

On balance, the space program has paid for itself through the creation of new technologies and scientific discoveries.

	Agree (%)	**Unsure (%)**	**Disagree (%)**
Space attentives:	76	3	21
Other citizens:	54	7	39

The American space program should try to land astronauts on Mars in the next 25 years.

	Agree (%)	**Unsure (%)**	**Disagree (%)**
Space attentives:	70	2	28
Other citizens:	50	6	43

In other words, even those not interested in space believe—by relatively slight margins—that our space program has paid for itself and that we should send people to Mars. And a 1988 *Time* magazine poll reported that 72 percent of Americans believe that a joint U.S.-Soviet Mars mission is a good idea. Oddly enough, support seems to grow when times are bad. Immediately after the *Challenger* accident, polls showed a strong surge of support for continued exploration of space.

No politician is going to come running to NASA with a basketful of money based on these results, but neither is anyone going to be voted out of office on that issue alone. Naturally those representing districts that prosper from the aerospace industry tend more toward a favorable view than do those from rural, agricultural areas. Also, polls show that men are a lot more apt to approve of the space program than are women. But there is solid support in this country for space exploration. Generally speaking it is unorganized support, but there are a few exceptions. A pro-space political action committee called Spacepac has existed since 1982, but it has been largely ineffectual. More recently Spacecause has been formed as the lobbying arm of the National Space Society. More effective than either of these is the Planetary Society, a 125,000-member group founded by Carl Sagan and Bruce Murray, the former director of the Jet Propulsion Laboratory, run jointly by NASA and Caltech. The society's able executive director is Louis Friedman, an astronautical engineer and writer who is also wise in the ways of Washington.

The Planetary Society has made Mars the centerpiece of its efforts, and no voice in this country has been more influential in space matters than Carl Sagan's. Wernher von Braun used to be the nation's chief space advocate, and NASA's unofficial spokesman, but his was a voice from within the agency and he emphasized rockets. Sagan represents science, and as an outsider he can set his own agenda and give NASA hell if he chooses. Bruce Murray, a geologist, is now a professor at Caltech and has close ties with the Soviet space science community. Sagan, Murray, and Friedman have put together the following Mars Declaration for the Planetary Society:

The Mars Declaration

> Mars is the world next door, the nearest planet on which human explorers could safely land. Although it is sometimes as warm as a New England October, Mars is a chilly place, so cold that some of its thin carbon dioxide atmosphere freezes out at the winter pole. There are pink skies, fields of boulders, sand dunes, vast extinct volcanoes that dwarf anything on Earth, a great canyon that would cross most of the United

States, sandstorms that sometimes reach half the speed of sound, strange bright and dark markings on the surface, hundreds of ancient river valleys, mountains shaped like pyramids and many other mysteries.

Mars is a storehouse of scientific information—important in its own right but also for the light it may cast on the origins of life and on safeguarding the environment of the Earth. If Mars once had abundant liquid water, what happened to it? How did a once Earthlike world become so parched, frigid and comparatively airless? Is there something important on Mars that we need to know about our own fragile world?

The prospect of human exploration of Mars is ecumenical—remarkable for the diversity of supporting opinion it embraces. It is being advocated on many grounds:

- As a potential scientific bonanza—for example, on climatic change, on the search for present or past life, on the understanding of enigmatic Martian landforms, and on the application of new knowledge to understanding our own planet
- As a means, through robotic precursor and support missions to Mars, of reviving a stagnant U.S. planetary program
- As providing a coherent focus and sense of purpose to a dispirited NASA for many future research and development activities on an appropriate timescale and with affordable costs
- As giving a crisp and unambiguous purpose to the U.S. space station—needed for in-orbit assembly of the interplanetary transfer vehicle or vehicles, and for study of long-duration life support for space travelers

- As the next great human adventure, able to excite and inspire people of all ages the world over
- As an aperture to enhanced national prestige and technological development
- As a realistic and possible unique opportunity for the United States and the Soviet Union to work together in the spotlight of world public opinion, and with other nations, on behalf of the human species
- As a model and stimulant for mutually advantageous U.S./Soviet cooperation here on Earth
- As a means for economic reconversion of the aerospace industry if and when massive reductions in strategic weapons—long promised by the United States and the Soviet Union—are implemented
- As a worthy application of the traditional military virtues of organization and valor to great expeditions of discovery
- As a step toward the long-term objective of establishing humanity as a multi-planet species
- Or simply as the obvious response to a deeply felt perception of the future calling.

Advances in technology now make feasible a systematic process of exploration and discovery on the planet Mars—beginning with robot roving vehicles and sample return missions and culminating in the first footfall of human beings on another planet. The cost would be no greater than that of a single major strategic weapons system, and if shared among two or more nations, the cost to each nation would be still less. No major additional technological advances seem to be required, and the step from today to the first landing of humans on Mars appears to be technologically easier than the step from President John F. Kennedy's announcement of the Apollo program on May 25, 1961, to the first landing of humans on the Moon on July 20, 1969.

We represent a wide diversity of backgrounds in the fields of science, technology, religion, the arts, politics and government. Few of us adhere to every one of the arguments listed above, but we share a common vision of Mars as a historic, constructive objective for the technological ambitions of the human species over the next few decades.

We endorse the goal of human exploration of Mars and urge that initial steps toward its implementation be taken throughout the world.

To date, approximately 150,000 people have signed the Mars Declaration. In addition to those with a vested interest in space, the list includes many more whose names do not usually appear on such lists: Jimmy Carter, former president of the United States; Roger L. Shinn, professor emeritus of social ethics, Union Theological Seminary; Rabbi Alexander M. Schindler, president, Union of American Hebrew Congregations; Barry Mano, president, National Association of Sports Officials; Jack Anderson, syndicated columnist; Lucy Wilson Benson, former national president, League of Women Voters; Tom Bradley, mayor, Los Angeles, California; Henry Cisneros, mayor, San Antonio, Texas; Susan Eisenhower, president, Eisenhower World Affairs Institute; Murray Gell-Mann, Nobel laureate, physics; Jerome Grossman, president, Council for a Livable World; the late Jim Henson, creator of the Muppets; Quincy Jones, composer; James Earl Jones, actor; the late Louis L'Amour, author; Eleanor Holmes Norton, former chair, U.S. Equal Employment Opportunities Commission; Joseph V. Paterno, head football coach, Pennsylvania State University; Linus Pauling, Nobel laureate, peace; the Reverend Norman Vincent Peale, author; Ester Peterson, consumer advocate; Russell W. Peterson, former chairman, U. S. Council on Environmental Quality; Sidney Poitier, actor and director; Ted Turner, president and CEO, Turner Broadcasting System; S. Dillon Ripley II, secretary emeritus, Smithsonian Institution; and Clyde W. Tombaugh, discoverer of Pluto.

Carl Sagan's great hope is to divert money from armaments to a Mars project. He particularly dislikes the Strategic Defense Initiative, or Star Wars, because he thinks the idea can't be made to work and that it will destabilize the uneasy nuclear impasse between the United States and the Soviet Union. The money being contemplated for SDI would finance a Mars expedition several times over.

A new president always gets plenty of advice, and waiting for George Bush when he took office were at least two studies on our future in space. The first, conducted by the National Academies of Science and Engineering, concluded that human exploration of the Solar System—without specifying Moon or Mars next—should be the guiding principle behind more immediate goals, such as the space station. The second space-policy study was prepared by the Center for Strategic and International Studies, a Washington think tank with a reputation for unbiased work. I was a member of this panel, which was cochaired by Brent Scowcroft, now President Bush's national security adviser, and Dr. John McElroy, dean of engineering at the University of Texas. Our report covered a wide range, urging White House leadership in education, the environment, competitiveness, and exploration. We concluded that the human exploration of Mars should be our long-term space goal, one that should follow a series of negotiated steps with our allies and with the Soviet Union.

The news media, especially since the *Challenger* accident, have been quite outspoken in urging a more coherent space policy with more clearly defined goals. The *Los Angeles Times*—not surprisingly considering its location in aerospace-rich southern California—is a leader in the pro-Mars contingent, but the *New York Times* seems equally enthusiastic. A host of other newspapers and magazines, including the *New Republic*, approve of the idea. The *Los Angeles Times* minces no words:

> The space program remains the most ambitious, daring and far-reaching effort of our age, and it should not be allowed to drift into mediocrity and complacency. The space program needs to recapture the excitement and imagination of the 1960s, when a lunar landing was its goal. A Mars landing is the way to do it. It is an important and worthwhile endeavor in its own right. . . . The question is not whether we can afford to do it. The question is whether we can afford not to. On to Mars.

Norfolk's *Virginian-Pilot*, my local newspaper when I am on the Outer Banks of North Carolina, is intrigued by the international aspects of a Mars mission:

> Plodding Russians, orbiting in space in reliable, if prosaic, hardware, and dashing Americans, leaping into space in fits and starts with the aid of exotic technology, might prove to be incompatible partners for a Mars

> venture. But they also might form a winning combination of caution and daring. Washington and Moscow should examine the pros and cons of linking hands again, as they did in the 1970s, in a space enterprise that could be most productive, both on Earth and beyond.

Short of a vote, it's hard to judge how Congress reacts to all this. There are congressional leaders who have supported a Mars initiative (though most insist it must come first from the White House). Hawaii's late Senator Spark Matsunaga, a Democrat, wrote an entire book on the subject, *The Mars Project: Journeys Beyond the Cold War.* Another Democrat, Senator Albert Gore of Tennessee, has visited Antarctica and written eloquently of the place and the science being performed there. I listened to him speak a couple of years ago, and he seemed favorably disposed toward an international Mars expedition, once the government has decided that it is, as he put it, "serious about space." Senator Gore now chairs one of the subcommittees that will decide NASA's fate. On the other side of the Hill, in the House of Representatives, there are key supporters on the pertinent committees. Congressman George Brown of California is one, Bill Nelson of Florida another. Representative Nelson and Senator Jake Garn are the only two politicians to have flown in space while in office, on board the shuttle. Nelson has written Vice President Dan Quayle: "As you know, I personally view a manned Mars mission as a critical long-term goal. This is an achievable, and indeed an inevitable, goal." There are others, such as Newt Gingrich of Georgia, who are avid space enthusiasts and will support expansion of NASA's role and budget. Gingrich writes enthusiastically of a huge commercial market in space: "Space is the 'Go West Young Man' of Horace Greeley in the 1870s. Space is the Atlantic Ocean of the seventeenth-century pilgrims. . . . Space can bring our nation flocking back to study math and science, eager and excited about learning enough to get out there."

One reason Bill Nelson lobbied Dan Quayle is that the National Space Council has been resurrected recently and the vice president chairs it. This organization, moribund since the Nixon days, will be the focus for hammering out the conflicting views of various departments within the executive branch. Beyond that, how much clout it will have with the president or Congress remains to be seen. As a senator, Quayle dealt with military space matters more than with NASA's problems, but he is said to be a quick study and apparently takes very seriously his duties as chairman of the National Space Council.

Which brings us to President George Bush himself. The Republican platform upon which he ran contains this amazing plank: "We must commit to a manned flight to Mars around the year 2000." Political platform statements are a place for excesses and for platitudes, and are meant to be taken with a grain of salt, but still—Mars by the year 2000? Some minor party functionary, if not Bush himself, really went out on a limb to include this provision (written by the Committee on Resolution at the Republican National Convention in New Orleans in 1988).

During the campaign George Bush mentioned a Mars mission favorably, but only obliquely. In a speech in NASA country—Huntsville, Alabama—he seemed most taken with a different one of Sally Ride's four options, Mission to Planet Earth: "Let us remember as we chase our dreams into the stars that our first responsibility is to our Earth . . . let us first preserve the fragile and precious world we inhabit."

Later, in a California speech, after having watched the shuttle land, candidate Bush said, "I am fully and utterly committed to the U.S. space program. . . . One thing we must do now . . . is develop a new and comprehensive strategy. . . . I am not completely sure Mars is the next place we ought to go, and I want to receive the best thinking on that." He then went on to express his preference for sending people into space rather than robots: "Men and women do not follow machines; they follow great men and women." Shortly after the election, Bush (according to *Aviation Week and Space Technology* magazine, a very reliable source) said, "The way I see it, the logical order is first the Moon, then—perhaps Mars."

As the new president, Bush inherited from the Reagan administration a 1988 presidential directive on national space policy. It includes as a goal "to expand human presence and activity beyond Earth orbit into the Solar System." That is about as specific as it gets in regard to Mars. NASA is directed to establish a technology program called Pathfinder to lay the groundwork for "a Presidential decision on a focused program of manned exploration of the Solar System."

Having talked to George Bush briefly on various occasions over the past twenty years, I feel that he is genuinely more attuned to the exploring spirit of America—and to our space program—than were any of his predecessors since Lyndon Johnson. But now he is faced with the enormous backbreaking reality of our national debt. That is the parent, and its child is the recent inability of our government to construct an annual spending plan that does not add to the deficit. It is easy for a presidential candidate to say, as George Bush did back in 1980, that "space funding is, in my view, a relatively low-risk investment capital on which the nation will realize a high return." But then, in the same press release, he added almost immediately: "The bottom line in our economy is that we have to balance the budget, *now.*" How to reconcile the two viewpoints?

The first clue came with his 1990 budget. Less than three weeks into his presidency, George Bush addressed a joint session of Congress and outlined a comprehensive agenda for "building a better America." In the text of his speech, under the heading "Advancing Priorities for Growth and Competitiveness," is a section on "maintaining America's leadership in space." It is both general and specific. It begins with a bit of philosophy. "In very basic ways, our exploration of space defines us as a people—our willingness to take great risks for great rewards, to challenge the unknown, to reach beyond ourselves, to strive for knowledge and innovation and growth. Our commitment to leadership in space is symbolic of the role we seek in the world." Next is a list of specific programs, with emphasis on the space station *Freedom.* Neither Mars nor a return to the Moon is mentioned. Finally, the president put some teeth in his proposal by asking Congress for a 22 percent increase in NASA's appropriation for 1990, from $10.9 billion to $13.3 billion. In today's climate of Gramm-Rudman-Hollings ceilings, a 22 percent increase shows a very strong presidential endorsement of NASA. (Unfortunately, the space station will consume nearly all of the increase.)

The twentieth anniversary of the first lunar landing got a lot of publicity and was a cause for much celebration among space buffs. President Bush's advisers considered it an opportunity he could not pass up, and he responded by making a speech at the National Air and Space Museum on July 20, 1989. Neil Armstrong, Buzz Aldrin, and I were on the podium with the president, the vice president, and their wives. It was a special time for me, looking back at those amazing Apollo years, but more than that I had a feeling of anticipation. Two weeks earlier I had been to a meeting with Vice President Quayle, National Space Council staff, NASA administrator Richard Truly, and a few outsiders. The tone was very upbeat: a return to the Moon, then on to Mars—damn the torpedoes, full speed ahead. But, I wondered, what would happen when these true believers got out into the real world with the guys with the green eyeshades and the cold eyes? Was there money for any of this stuff?

George Bush's speech was relatively short, but it was a major pronouncement nonetheless. He reminisced about the good old days, and then he looked to the future: "Today the U.S. is the richest nation on Earth—with the most powerful economy in the world. And our goal is nothing less than to establish the United States as the preeminent spacefaring nation." To do this, he said, "I'm not proposing a ten-year plan like Apollo. I'm proposing a long-range, continuing commitment . . . first . . . for the 1990s . . . space station *Freedom* . . . next—for the new century—back to the Moon . . . back to stay. And then . . . a manned mission to Mars." He went on to say that one purpose of *Freedom* was to pursue Mission to Planet Earth, an international initiative to seek solutions for ozone depletion, global warming, and acid rain. Beyond that, "Today I am asking Vice President Quayle to lead the National Space Council in determining specifically what's needed for the next round of exploration—the necessary money, manpower, and material—the feasibility of international cooperation—and [to] develop realistic timetables and milestones along the way." With that, we all moved over to the White House, where President and Mrs. Bush hosted a barbecue luncheon for former astronauts and space officials.

Munching on a hot dog, I tried to digest what the president had said—and not said. Certainly NASA had to be pleased, because he had endorsed everything on the agency's agenda. *Freedom*, Moon, Mars—it was all there. But what was missing might be even more important: some indication of how high a priority his administration would put on space when the inevitable budget crunch came. Also the speech had a weak ending: Vice President Quayle was to study the matter and report back. George Bush's peek into the twenty-first century was not as dramatic or aggressive as John F. Kennedy's demand for the Moon, yet if the Bush agenda is actually followed, it will dwarf the accomplishments of Apollo. A permanent colony on the Moon and a mission to Mars: extraterrestrial settlements will become part of our civilization.

After the speech, administration officials filled in some details. The vice president's recommendations were expected in six to nine months. NASA indicated that a Mars landing might be possible by the year 2016. Budget Director Richard G. Darman estimated that the Moon-Mars program could cost $400 billion over thirty years, or $13 billion per year. Congressman Bill Nelson, chairman of the House Space Subcommittee, predicted that NASA's appropriation would have to exceed $30 billion per year to do everything Bush wanted. Reaction on Capitol Hill and elsewhere was generally favorable to the president's ideas but critical of his budget planning, or lack of it. "There's no such thing as a free launch," House Majority Leader Richard A. Gephardt said. Senate Budget Committee Chairman Jim Sasser accused the president of making a "giant leap for starry-eyed political rhetoric." "A daydream as splashy as a George Lucas movie, with about as much connection to reality," snorted Senator Albert Gore. Congressman Leon Panetta: "You can't go to Mars on a credit card." On the other hand, John Logsden of George Washington University, a space-policy analyst and expert on the Kennedy days, said the speech was "exactly the right thing to say. He's set a vision, set goals, and challenged the country to face up to whether they want to do it or not."

No surprises in any of this, but wait a minute: it turns out that the guy with the green eyeshade is also an unabashed space buff! Richard Darman, who is responsible for putting together President Bush's budget, made a speech on July 20 just hours after his boss did. He said he considers the space program "the preeminent symbol of public policy commitment to the future." And we need to pay more attention to the future, Darman believes, to combat what he calls now-now-ism, a "shorthand label for our collective shortsightedness, our obsession with the here and now." This myopia, according to Darman, has resulted in drug abuse, a decline in education, and economic practices that favor spending over saving. "The American Dream," he concludes, "is not meant to be filtered through green eyeshades." This is an unusually positive speech for a Washington budgeteer, and it makes President Bush's lofty space goals seem a little closer to realization. Maybe Darman *can* find the money.

Media reaction to the president's speech was mixed, but

the prevalent opinion seemed to be that of George Johnson, writing in the *New York Times:* "Lacking a timetable or a budget request, Mr. Bush's promise that the country would build a lunar base and send people to Mars sometime next century struck many people as abstract and unconvincing." There is no doubt that opposition to a Mars project will be formidable. Opposition may be more difficult than support to quantify, because there are no antiplanetary societies, and there are plenty of more immediate targets, such as SDI. But although they don't write many letters to the editor about Mars, millions of Americans undoubtedly feel (1) it is too expensive, (2) there are too many problems on Earth to be solved first, (3) there's nothing in it for them, and (4) they could care less about the place (if it even exists). In this view, space advocates are "dreamy-faced loons," according to *Washington Post* columnist Henry Mitchell, whose writing I usually enjoy. Among opposition government officials, former Senator William Proxmire sticks out in my mind: he used to love to give NASA his well-publicized Golden Fleece Award for wasting tax dollars.

Even among the dreamy-faced loons there is opposition to Mars. A model of consistency, physicist James Van Allen is always against new manned operations: "Any serious talk of a manned Mars mission at this time is grossly inappropriate," he has said, although he does concede, "it's a matter of high adventure." Then there are the "Moon-first" people, who don't want to discuss Mars until after elaborate plans have been agreed upon to set up some kind of astronomical observatory or prototype colony or manufacturing base on the Moon. Former NASA administrator James Fletcher seems to be in this camp, as does Sally Ride. There is also squabbling among the Martians. Usually the objection is to including the Soviets, because they will steal all our secrets—technology transfer, it is called. Experts, usually military ones, point to the Soviet shuttle *Buran* as a case in point. (Of course the Soviets obtained that technology without a joint Mars mission, but their shuttle didn't launch until eight years after ours.) Others think that the Soviets are unreliable partners for technical or political reasons. The recent failure of both Phobos probes is cited as evidence of technical incompetence. Politically our history of bilateral relations has been a model of inconsistency. Americans and Soviets don't seem to agree on anything for very long, certainly not long enough to carry out a joint mission to Mars. Space writer Alcestis Oberg sums up this view: "A joint mission completely and utterly ignores reality . . . the potential for spying, for technological transfer, for interference in our political system, for the 'hostage holding' effect it would have on our space program and our future." Still others cite esoteric technical difficulties, such as designing equipment to each other's specifications and the problem of reaching the same Earth orbit when launching from as far north as Baikonur and as far south as Cape Canaveral.

It is in this environment that President Bush must reply to the Soviet invitation of 1987 to explore Mars jointly—or perhaps he can ignore it. One difficulty in simply accepting is that the United States might seem to lack vision and ideas of its own, as if it were jumping onto the Soviet bandwagon—although the president's July 20 speech dispelled that notion somewhat. A better response, in my view, would be a counterproposal to the world at large to join the United States and the Soviet Union in a long-range program of Mars colonization. As long as the idea remains far off in the twenty-first century, the U.S. government can get by with just words and no money. But the camel's nose would be under the tent flap. Funding for Martian preliminaries would have a slight impact on NASA's budget during George Bush's first term and a major effect during his second term, if there is one.

The National Space Society, labeling current U.S. space efforts "inadequate," has called on the nation to initiate a Decade of Doing rather than simply conducting more NASA studies. George Bush's stirring speech of July 20 may well usher in such a decade, but much political groundwork must be done before that can happen. I can't get out of my head the story that Christopher Columbus had to wait seven years just to get an appointment with Queen Isabella.

CHAPTER

XIX

SOME PRELIMINARIES

NOW THAT President Bush has roughed in the broad outlines of the next thirty years in space, I expect a long-term debate on if, when, and how it will all get done. The dreamy-faced loons will have to defend their position using every political, economic, and scientific argument, including negative and chauvinistic ones ("The world will pass us by if we don't"). Money, of course, will be the problem (although it will never get any cheaper); some space supporters may agree to put their dreams on hold in deference to the national debt. Still others will be fearful of bringing the entire package to a vote in Congress lest they lose and set the cause back indefinitely.

But there are also signs of great hope. If Mikhail Gorbachev prevails and substantially reduces Soviet defense expenditures, then surely this country will follow suit. Apparently, after half a century we are going to bring some troops home from Europe. The money saved might be diverted to SDI, but I doubt it, because as time goes by I believe the SDI strategy increasingly will be seen as flawed. There are just too many counters to SDI, not the least of which would be for Soviet planners simply to bypass our impregnable shield overhead and concentrate on the soft underbelly by strengthening their low-altitude attack capability. It is more likely that, as in the case of our Western allies, who are more and more taken with Gorbachev, the United States citizenry will welcome a substantial reduction in our armaments budget, including SDI. They may not welcome the resultant unemployment in the defense industry, at present a $300 billion per year enterprise touching all fifty states. And make no mistake: the shift from guns to butter will be painful for some. Every time the Pentagon sends Congress a list of bases it wants to close, the local representative suddenly drops his or her latest waste, fraud, and abuse speech and launches an eloquent defense of Base X and its contribution to the regional economy. Defense plant closings hurt even more, but if NASA and Mars could absorb some of their work, the local economies could glide to a soft landing before taking off again on a different heading. If NASA could get its hands on just 10 percent of the Defense Department's money, the NASA budget would be restored to its buying power in 1966, the peak of the Apollo years. Experts agree this would be more than enough to fund everything George Bush called for on July 20, 1989.

At the same time NASA would serve the useful purpose here on Earth of easing the transition of a vital, high-tech industry from military to civilian endeavors. This transition is important from several viewpoints. First, a Mars venture would maintain our national technology at a high level, something that cannot be accomplished by diverting the funds into low-income housing, agricultural subsidies, or highway construction. Modern technology is essential for our national security and for our manufactured goods to compete on the world market. On the one hand, sharing our technology with the Soviet Union on a joint Mars project would be no loss because it is available anyway in our unclassified literature. On the other hand, we would be more aware of Soviet capabilities and perhaps intentions. As in the case of reconnaissance satellites, knowing what the other side is doing is generally stabilizing, and therefore good. In a similar fashion, collaboration with the Japanese and Europeans would probably help, not hurt, our chances of competing successfully in the commercial marketplace.

This rosy scenario may take a few years to develop, but there are things that can be done now to help it along. I leave national security matters to others, but just within NASA's small world there are low-cost options that could materially influence U.S-Soviet relations. First is a friendly response to Gorbachev's invitation to explore Mars jointly. Second is some manifestation of celestial cooperation, perhaps by sending the U.S. shuttle to visit the Soviet *Mir* space station. There are technical difficulties in doing this, and it might not be possible to dock the two, as the Apollo and Soyuz rocket did in 1975. But a joint flight would be a symbolic gesture of great importance, and a photograph of the shuttle and the *Mir* parked side by side in Earth orbit (taken from a Soyuz) would be a powerful talisman for future cooperation.

Then there are unmanned Mars flights already planned by the two nations. Despite their two Phobos failures, the Soviets apparently are trying to stick to their original plan, which called for a 1994 launch of a mission that would deploy one or two balloons in the Martian atmosphere, drop some small weather stations, and penetrate the surface with several probes. But now they are having trouble defining the details of the mission, which includes French and American participation. For example, Pasadena's Planetary Society is designing the payload for one of the balloons, a snake-shaped apparatus that will drag on the ground at night, when the balloon cools and descends to near the surface. The U. S. government should encourage this kind of participation.

On the U.S. side, we are planning a Mars Observer flight to be launched in 1992. This probe will be relatively simple. Its goal is to study Mars's climate and surface for one Martian year (687 Earth days). From polar orbit, it will map the entire Martian surface twelve times, study dust storms and their formation, measure the gravitational field of the planet, and provide some data on the chemical composition of its surface. Compared with the two Viking soft landers of 1976, Mars Observer will be something of an anticlimax, but it will assist designers who want to know more about Martian clouds, storms, frost, wind velocity, and soil chemistry. Mars Observer will also move us closer to selecting a site for the first human colony. American space officials have agreed to install a special radio receiver on the Mars Observer to handle transmissions from the Soviet balloons on board their 1994 flight. The Soviets should also be invited to participate to the maximum extent in analyzing the data returned by the Mars Observer.

The year 1992 will be significant in other ways, too. In 1985 Senator Spark Matsunaga proposed that 1992, the five-hundredth anniversary of Columbus's voyage and the seventy-fifth anniversary of the Russian Revolution, be designated as the International Space Year. His resolution was endorsed by both houses of Congress and signed by President Reagan. From this grew an international organization called SAFISY—Space Agency Forum for the International Space Year. More than twenty nations belong to SAFISY, and they are working to make 1992 the first yearlong, worldwide celebration of humanity's future in space. SAFISY is also establishing panels to discuss specific topics. There will be meetings in Japan in 1990, in the Soviet Union in 1991, and in the United States in 1992. SAFISY strikes me as a ready-made forum for discussing an international Mars venture, and for considering the immense problem of organizing such a complex undertaking. The most likely solution may well be an equal partnership between the United States and the Soviet Union, with other nations acting as subcontractors to one or both of the superpowers. But there may be better ways of doing it, and with an apparatus already in place, SAFISY could be the mechanism for organizing it. After 1992 this panel could be disbanded along with the rest of SAFISY, or it could remain active until a sturdy skeleton organization had been created.

One of the first tasks of a Mars organization would be to assemble a master plan of the steps required: research and development to fill any gaps in the technology; robotic precursor missions; manned dress rehearsals in Earth orbit, if required, or on the Moon; the first Mars landing; follow-up landings; the beginning of a colony. These would be the required steps, although I may have left some out. The first step, defining what we need to know but don't, is of fundamental importance. In a 1987 speech Jim Fletcher, then NASA administrator, ticked off seven areas we need to know more about for a successful Mars venture. (1) the life sciences, (2) controlled environmental life-support systems, (3) the design of better and cheaper rockets, (4) space medicine and surgery, (5) aerobraking, (6) the storage and handling of cryogenic fluids, and (7) large-scale assembly and servicing of machinery at a station in Earth orbit. Our fledgling Mars organization needs to come to grips with the items on this list—or to devise its own list—and needs to incorporate all this learning into its master plan. At present the United States is doing research in all these areas, but some of it does not enjoy a high priority. The shuttle and the space station don't leave much room in the NASA budget for anything else, although a technology improvement program called Pathfinder has been approved. I assume that the Soviet Union is having similar funding difficulties, especially since the two Phobos failures have—in the spirit of *glasnost*—stirred up a lot of opposition to space spending.

A good beginning for the two superpowers would be to apportion these research tasks, either between themselves or by assigning them to other participating nations. The most difficult one, in my opinion, is creating a reliable closed life-support system. The tidy engineer in me says that this is apt to be an icky, gooey mess. It's not sliding metal parts; it's foul odors and slimy substances. Throwing money at it will help, but beyond that a lot of brainpower will be needed. This is a crucial area of development, and one that the United States and the Soviet Union probably will decide to work on simultaneously, sharing results but with each country creating its own hardware. There is safety in redundancy. So far the Soviets have done more work in this area than we have: aboard *Mir* they recycle water condensed from cabin air; we have always dumped wastewater overboard. Cooperation therefore might entail the transfer of technology from them to us.

I sometimes wonder whether the Soviets have as large a contingent concerned about technology transfer as we have. I don't worry too much about the technology transfer implications of a Mars flight because the Soviets have so many other ways of gaining access to such information in this country, and because the Mars equipment, while it will be very sophisticated, will not be the kind that converts easily to military use. General Tom Stafford, commander of the Apollo that docked with Aleksei Leonov's Soyuz, told me that the main thing the Soviets could copy from that mission was our organization and management, but he felt that under their political system it would be very difficult to apply such lessons. At any rate an appropriate division of responsibility would eliminate, or at least reduce, technology transfer.

One omission from Dr. Fletcher's list of things we don't know enough about is whether there are dangerous microbes on Mars. From Viking results we are almost sure there are none, but no one can absolutely guarantee it. We had the same problem on *Apollo 11* in regard to Moon dust. No one worried quite so much about the Moon, however, because it is a much less hospitable place than Mars. The Moon has no atmosphere, and there are no credible scientific theories about life having once existed there. Nonetheless, Armstrong, Aldrin, and I were kept in quarantine in a Houston laboratory until August 10, 1969—three weeks to the day after the lunar landing.

In the case of Mars, I suppose it only prudent to examine a soil sample before risking humans. That examination will probably be done aboard a space station in Earth orbit. Independent of one another, the Soviet Union and the United States have each planned an unmanned sample return mission whose design will push robotic technology to its limits. Scientists want a roving vehicle that can roll (or walk?) some sixty miles away from a Martian base camp and return with a variety of samples, some of them bored from subsurface layers. One design concept is a seven-legged, collapsible T-beam that walks by sliding a central platform back and forth while four platform-mounted legs and three T-beam legs alternately retract and extend. People on the Earth will be guiding the rover, but with a communications delay of as long as twenty minutes it must be nearly autonomous in its ability to cope with difficulties—able to bridge crevasses and step over obstacles without getting bogged down in the sandy Martian soil. Furthermore it must be programmed to expect the unexpected, to measure and record anything startling. But what should it consider out of the ordinary? An object of a particular size, or some unusual color, or shape, or movement, or what?

Timetables vary, but the year generally mentioned for this flight is 1998. Merging the two countries' efforts would certainly save money, might save some time, and would be an ideal test for our Mars organization. To minimize technology transfer, one side could be responsible for getting the roving vehicle to Mars, and the other for getting the sample from the rover back to Earth orbit. Thus the hardware would be divided into two tidy packages that would have to fit together only at the point where the sample was transferred from the rover to the return vehicle. Similar suggestions have been made in regard to a manned flight: each side aims for the same spot on Mars, meets there, conducts joint surface operations, and returns to Earth independently. I don't like that idea because it's hardly a joint mission at all and would be more expensive. But a manned mission could certainly be divided into national modules. How many of them, and how much redundancy, depends on a trade-off between safety on one hand and weight and cost on the other. In my proposed profile, I suggest two nearly identical round-trip modules but only one of everything else.

Another open question is the number of preliminary flights that will be required before attempting to land humans. I don't know whether the Apollo experience will apply in this case, but we found that after a slow start, once we began flying, things went quickly. *Apollo 11*, the first lunar landing, was preceded by only four manned flights, although early planning charts had shown as many as ten. One big decision for Mars planners will be how to use the Moon as a stepping-stone. Earlier in this book I discussed Antarctica and space station *Freedom* as locations for testing Mars equipment and crews. Many experts, not to mention George Bush and Dan Quayle, think the Moon should be added to this list because conditions there are somewhat Mars-like yet it is only three days from home in the event of trouble. People on the Moon would have to live inside domes, supplying their own atmosphere. They would have to be protected from solar flares. The Moon can be colder than Mars as well as a lot hotter. The Moon's gravity field (one-sixth Earth g) is fairly close to that of Mars. The Moon is isolated from Earth, psychologically as well as physically. Other people go a step further and say that the Moon's resources can be mined and used as rocket fuel or as building materials for a Mars craft. They point out that a very expensive component of a Mars mission is hauling fuel up from the Earth's surface, fighting Earth's strong gravity. Fuels produced on the Moon would not suffer this penalty and would be cheaper, or so the argument goes.

I reject the whole Moon idea. I think it is an unnecessary detour on the road to Mars, and one that could gobble up the entire Mars budget. A rocket-fuel factory, whether on the Moon or on Phobos, would be extraordinarily expensive and should wait, in my judgment, until we have demonstrated we can exist successfully for an indefinite period on either planet. Mars is a far more interesting place than the Moon, although the Moon may eventually be mined with commercial success. In the meantime the safety aspects of a Mars program need to be examined carefully. I don't think the "three days from home" argument holds up. If we can't trust equipment beyond that range, then we should be working out those bugs even closer to home, in Earth orbit or Antarctica or some laboratory in California. A Moon base might even teach us some bad habits by letting us be more dependent on Earth supplies than would be possible on a Mars colony. I could be wrong, but I don't think the American public will be at all enthusiastic about a return to the Moon. The Moon? Didn't we already go there?

Another fundamental decision is whether or not a Mars ship needs artificial gravity. It is to be avoided if possible because it adds another engineering problem to a machine that will be sufficiently complicated without it. The decision depends on how the medical community—Soviet and American—analyzes the data from past and future long-duration flights. I think the one-year flight on board *Mir* proves that gravity is not needed, but that may be a reckless view in the eyes of many experts—including some cosmonauts and astronauts.

Artificial gravity can be created by rotating a spacecraft to produce centrifugal force. The engineering problems involve starting and stopping this rotation, and keeping it turning smoothly. Sometimes on *Apollo 11* we were required to rotate slowly, like a chicken on a barbecue spit, to distribute the sun's heat evenly and prevent pipes from freezing or boiling. Unless we executed the spinning maneuver with the utmost precision, we found that we would begin to nutate (wobble like a top), and then we would have to stop and start all over again, wasting more fuel in the process. A Mars machine will be a lot more sophisticated but fuel usage will still be a factor. Also, while turning it is more difficult to perform navigational chores or maintain communications. There may also be physiological problems if the radius of rotation is not sufficiently large, perhaps a hundred feet or so. Rotating creates something called Coriolis force, a swirling acceleration that disturbs the human vestibular system and can cause loss of balance, nausea, and disorientation. To avoid these difficulties, the most popular scheme is to connect two spacecraft by a long tether, or tunnel. After starting a rotation, centrifugal force would keep a tether taut. I say artificial gravity is not needed, but we have plenty of time to find out, using *Mir* and *Freedom* as test beds. If necessary, the two Mars round-trip modules could be so connected and rotated.

Going to Mars would certainly impose an extra burden on NASA, but in my opinion it would also be the salvation of the agency. Ever since the end of Project Apollo NASA has been running out of steam. *Skylab* was a legacy of the Apollo program, as was Apollo-Soyuz. The shuttle required some brilliant engineering but it has been a disappointment in many ways apart from the death of seven *Challenger* crew members. The space station may excite some space buffs, but generally it has been greeted with a great yawn, perhaps because the Soviets have had one for such a long time. A return to the Moon might rejuvenate NASA, but personally I hope that Mars can come first.

NASA needs a new place, Mars—and needs it badly. It's not just the money that NASA needs, it's having a goal, a unifying vision for the future. I joke with my NASA friends that the agency should be renamed NAMA, the National Aeronautics and Mars Administration. NAMA would certainly have a clearer, if narrower, focus than NASA. NAMA would immediately understand why a space station is necessary and would design it differently. Instead of attempting to make it be all things to all users, the station's designers would tailor it to the needs of a Mars mission: a place to test concepts, people, and hardware. In a climate of healthy expansion, a Mars project would build not just Mars hardware but a creativity and enthusiasm that would spread to all facets of the space agency's work. For example, during the peak Apollo years, support for the space sciences was also robust, and some of the most productive unmanned flights were launched along with the Apollo missions. A Mars rocket would pull along an assortment of other ideas and projects in its wake.

Today NASA is having difficulty replacing its aging engineering work force with equivalent young talent. NAMA would have no such problem. Even though government salaries are low by industry standards, I believe NAMA could attract the brightest and the best new graduates, all eager to participate in a bold and challenging national or international push toward Mars. A healthy, vigorous space agency is a national treasure that deserves to be nurtured. And, sure, it could still be called NASA.

CHAPTER

XX

DESIGNING

IN PLANNING a Mars mission the United States and the Soviet Union should each design and build a round-trip module with nearly identical functions. By providing this redundancy, an equipment breakdown in one module would be an inconvenience but not a catastrophe. Each module will be conceived and constructed independently: a design flaw in one would not cripple both. Doing everything twice will obviously be more expensive, but I believe that, for safety's sake, duplication is required in equipment that must continue to operate over the entire twenty-two-month mission.

These two round-trip modules will be huge by present spacecraft standards because so much fuel, water, and other expendables must be carried. Aside from the problem of scale, however, the technology is well in hand except for one item, the controlled ecological life-support system, or CELSS. NASA's current research in this area does not enjoy a high priority. It must be accelerated unless we are willing to rely on the Soviet Union to design such a system, and I do not think that is wise. The Soviets are probably ahead of us now in CELSS development, but they also have a long way to go before a closed-loop system can be trusted for two years of faithful operation. Projects such as Biosphere II in this country and a Soviet experiment in Siberia may prove to be helpful, but any terrestrial enclosure can serve only as a rough guide to what might (or more important, might *not*) work in space. This country will use the shuttle to perform experiments, but the shuttle is severely limited in volume and in its capacity for extended flights. The Soviets are in a much better position with their *Mir* space station. We need access to their data, but we also need to design our own system to protect the fragile human bodies on board.

As far as CELSS is concerned, I think of the human body as a sealed box with one pipe in and one pipe out. In go oxygen, water, and food; out come solid and liquid wastes, germs, and carbon dioxide. The outlet pipe is fed into a second sealed box, the CELSS. It must be almost as magical as the first, for it must transform these by-products of the body into fresh supplies and pipe them back. A closed loop, easy to visualize but oh so difficult to keep working efficiently year after year. Yet a Mars mission cannot afford the weight penalty associated with throwing waste products overboard, as we have done in the past. Perhaps it is too difficult to recycle food and solid wastes, at least for the first Mars mission, but certainly the oxygen and water loops need to be closed. In other words, food may be grown on board both round-trip modules but only as supplemental, or luxury, items. The main diet will consist of stored food, and solid waste products will be dried and stored or thrown overboard.

Except for the thorny problem of CELSS, I believe the design of the round-trip modules will be relatively straightforward. The Soviets may use components from *Mir*, and we from space station *Freedom*. The hull will be aluminum and fabricated with conventional techniques. Wherever possible, highly reliable off-the-shelf equipment will be used. For propulsion I propose chemical rockets rather than advanced schemes. For electricity, solar cells and batteries will be too large, heavy, and unwieldy; a more compact nuclear electric power plant will be required. One currently in development, a 100-kilowatt system called the SP-100, will probably fit the bill. I assume the Soviets have equivalent devices because of their extensive research in this area and their use of nuclear power aboard military reconnaissance satellites.

If I were designing our round-trip module I would separate living quarters and workspace to create an ambiance that replicates our Earth-bound habit of cycling between labor and leisure. Some functions do overlap; for example, daily exercise can be considered work or recreation. I would make the exercise compartment an adjunct to the infirmary. Adjacent to that would be a science facility and the command and control center. A modest repair shop would complete the work area. I consider food preparation an off-duty activity, and so my galley would adjoin the wardroom, which is a combination living room, dining room, and library. Nearby would be the bathroom and sleep spaces for four. The Soviet round-trip module will house the other four crew members in whatever arrangement its designers prefer. Although the functions of the U.S. and Soviet modules would be nearly identical, each nation's engineers might arrive at different solutions to the same problems. The only components not duplicated will be those considered not essential to mission success. For example, conducting scientific experiments en route will be interesting, but not mandatory. Therefore the science facility in the Soviet module may specialize in medicine and physiology, while the American counterpart might be equipped for astronomical observations. Furthermore, it will be possible to interconnect the two modules in such a fashion that fluids (water and fuel) can be piped from a defective module over into its healthy counterpart. If future research shows artificial gravity to be necessary, the two modules will have to rotate, and the tunnel connecting them lengthened to more than fifty feet to minimize Coriolis disturbances. This would be an engineering nuisance but not an impossibility.

Except for the two round-trip modules, machines will not be duplicated. Too many special-purpose vehicles are required, in some cases to be used only for a few hours. The list includes a descent module to land four people on Mars, an unmanned lander carrying supplies and equipment, a surface habitat, a roving vehicle, an ascent and rendezvous module, and an Earth-capture vehicle to dock with an Earth-orbiting space station. En route to Mars, these modules will be attached to either the U.S. or Soviet mother ship and mounted on an aeroshell. The aeroshells, approximately one hundred feet in diameter and used for deceleration in Mars's atmosphere, will be jettisoned after use.

I expect that the manned descent and ascent stages will

be combined. Just as with the Apollo lunar module, the descent stage will serve as the launchpad for the ascent stage. Again like Apollo, I expect that each will have only one engine, although its thrust chamber will be fed by multiple pumps and valves, and other components will be duplicated wherever possible. Unlike the Moon, Mars possesses enough of an atmosphere to permit the descent vehicle to use aerobraking and parachutes in addition to a rocket engine, greatly reducing the amount of fuel the lander must carry.

For the first landing, both the habitat and the rover will be kept as simple as possible. Four people will spend forty days on the surface. Two at a time will explore with the rover. A nuclear power source will be set up some distance from base camp to minimize the harmful effects of its radiation. Solar radiation will be a larger problem, and the crew may have to partially bury the habitat to shield it from ultraviolet rays and other damaging particles from the Sun. When the four depart, the habitat and rover will be left behind for the next visitors.

Ascent and rendezvous should pose no new problems, but lives depend on its success, and there is something unsettling—to me at least—about the notion of having to find a tiny target in the vastness of space in order to get home safely. We did it flawlessly on Apollo but I still worry about it. Once docked with a round-trip module, the ascent stage will be abandoned, again continuing in the Apollo tradition. Approaching Earth, the crew of eight will transfer into the Earth-capture vehicle and will use its aeroshell to slow down enough to be captured by Earth's gravity. Then a sequence of rendezvous maneuvers will lead it to either a Soviet or U.S. space station, from which the crew will be returned to Earth in a shuttle.

These modules and all their equipment will be far too heavy to be lifted as one unit from the surface of the Earth. Even broken down into manageable pieces, they will still require the services of a huge launch vehicle, one about the size of the Saturn V that sent us to the Moon. Today the Soviets have such a booster, the Energiya. We do not, and even if we could re-create the Saturn V, we wouldn't want to; it would be far too expensive. A new, cheaper, big dumb booster is on the drawing board, under the watchful eye of the air force. NASA doesn't like this arrangement, but if SDI is ever deployed, the air force will have to lift equipment into orbit that will be far heavier than that required by a Mars expedition. NASA, mistrustful of its military rival, would rather modify its own shuttle system by retaining the two solid rocket motors and the external tank, but substituting an unmanned cargo carrier in place of the shuttle itself. NASA calls this the Shuttle C, and it could be flying long before the air force behemoth, known as the Advanced Launch System, or ALS. Either Shuttle C or the ALS would suffice for lifting the American components of a Mars mission into place, but the ALS would be a better choice because it should be considerably cheaper, on the basis of dollars per pound, to orbit. That is because it is being designed from scratch with new technology, and with the emphasis on economy rather than on performance.

Cheaper yet would be to use the Energiya, with its development costs behind it, to launch the American as well as the Soviet components, but I don't think that's in the cards.

For one thing, there is the problem of getting U.S. payloads over to the Soviet Union, or worse yet of transporting the Energiya to an American launchpad. But that's nothing compared with the hue and cry that would arise from the U.S. launch industry at losing all that business. There is a law prohibiting U.S. government payloads from flying on Soviet rockets, and it would be extremely difficult to get it changed. The mission profile that I propose is for the United States and the Soviet Union to launch their own components separately. That arrangement solves some problems but creates another one: getting everything together. Baikonur and Cape Canaveral are separated by eighteen degrees of latitude, which makes it uneconomical to deliver a payload from either launchpad to an orbit above the other—so much fuel would be wasted getting there that there wouldn't be much room left for payload. The same applies to the orbits of *Mir*, at 51 degrees inclination to the equator, and *Freedom*, which will be at 28 degrees. It will not be practical to assemble all the Mars hardware at either space station. What I propose is for the Soviets to do their assembly at *Mir* while we do ours at *Freedom*. One round-trip module, plus half of the other assorted vehicles, would be assembled at each place. They would set course for Mars simultaneously and join up about a week later, at a distance where the latitudes of their points of origin had become insignificant.

Assembly of the modules in Earth orbit will be a gargantuan operation and may require the development of new techniques. The U.S. shuttle has a manipulator arm that allows it to grapple cargo, and it has been used successfully both to deploy new satellites and to retrieve old ones. Astronauts have also taken space walks from the shuttle and repaired satellites. For that matter, I did a similar thing back in 1966, space walking to retrieve an experiment package from an Agena rocket and returning it to Earth aboard a Gemini spacecraft. So the idea of astronauts working in space and physically moving equipment is not new. The foundation has been laid, but compared with assembling a flotilla of Mars spacecraft what we've done so far has been like playing with Tinkertoys. First the aeroshell will have to be put together, because although it will be light, it will be too big to be launched on one rocket. Then the round-trip module will have to be assembled, followed by all the special-purpose vehicles. I estimate the task will require about ten deliveries (of 200,000 pounds each) by Energiya to *Mir*, and ten by ALS to *Freedom*—twenty huge packages of components to be joined together by methods yet to be developed. The pieces will have been built to interlock, but getting them all together still will be a formidable engineering feat. Certainly *Freedom* as currently conceived will not be an adequate facility for this kind of factory work. Probably a large hangarlike enclosure will have to be added to *Freedom* and similar modifications made to *Mir*. As much of the assembly as possible should be done indoors, in the relative safety of a pressurized compartment. Why run the risk of having astronauts perform intricate tasks while exposed to the vacuum of space? Pressure suits have only one thin rubber bladder, and if it tears or is punctured, the astronaut loses consciousness within seconds and is dead within minutes.

Although assembly will be at Soviet and U.S. space stations, the individual components will be manufactured in a number of other countries. The strong background and rich experience of the European Space Agency, and the emerging capability of the Japanese, will surely result in each producing at least one of the special-purpose vehicles. Perhaps one could be responsible for the manned descent/ascent stage while the other produced the surface habitat. Slightly smaller pieces, such as the unmanned supply landing craft and the roving vehicle, could be built by other capable countries such as Brazil and Canada. Such an arrangement would certainly relieve the Soviet Union and the United States of a large share of the financial burden, but it would add materially to the managerial complexity of the undertaking. At no place would this be more apparent than at one of the space stations during assembly operations: scheduling and carrying out deliveries on time, seeing that the pieces actually fit together properly. One glitch in Earth orbit and things on the ground would grind to a halt until the problem was fixed.

Under such conditions, when is it reasonable to set sail for Mars? Not until the second decade of the twenty-first century, according to NASA's interpretation of President Bush's mandate, with at least a decade devoted to establishing a lunar base first. But I remember that John F. Kennedy launched Project Apollo on May 25, 1961, and Neil Armstrong stepped out onto the Moon on July 20, 1969, eight years and two months later. A Mars venture will be a lot more complex than Apollo, but we know so much more today than we did in 1961. If the Moon is bypassed, I believe that a dozen years will suffice between go-ahead and the first Mars landing. Many at NASA judge this too optimistic a schedule, considering their preliminary task of launching space station *Freedom.* But others think that NASA is not receptive to innovative approaches. In late 1989 the National Space Council sought other opinions and invited a presentation by scientists from the Lawrence Livermore National Laboratories in California. By using inflatable structures made of the fabric Kevlar, and by bypassing *Freedom* altogether, the Livermore team believes they could reach the surface of Mars by the year 2000, at a cost of $10 billion.

At any rate, it is important to have a schedule with milestones and deadlines so that everyone involved knows what is expected and when. Without a firm timetable, it is only human to put things off. In some cases (CELSS for instance) hardware decisions will have to be postponed until more research has been done, but without a launch date to work toward, research will tend to become an endless process and costs will skyrocket. "By the end of the decade" was a wonderfully clear call for Apollo. Mars needs a similar deadline, one that is not so tight as to require shortcuts but one that is close enough to keep a little pressure on the participants. A launch in 2004 is, I believe, such a compromise. Will the Soviets and the other participating countries agree? I don't know, but the only way to find out is to propose it. To make the first Mars landing in 2005 is certainly possible, but it will require national resolve and a high degree of discipline from all participants.

CHAPTER

XXI

GETTING IT TOGETHER

THE EIGHT people chosen to make the first flight to Mars are probably around twenty-five years old today, most likely in graduate school studying engineering, science, or medicine. Four are men, four women. The United States will be responsible for selecting four of them, as will the Soviet Union. The selection process will be complicated by the decision to narrow the search to married couples, and by having two of the couples come from countries other than the United States or the Soviet Union. There will be a multiyear worldwide search for talent in the appropriate disciplines.

To make it easier for us to keep track of the eight during their training and twenty-two-month flight, I have given them specific identities and even names, the better to delineate individual responsibilities as we fly along with them to Mars and back.

In January 1994 NASA announced the selection process with much fanfare and, as expected, the response was immediate and overwhelming. Over ten thousand applications flooded in, but the vast majority failed to meet all the criteria. According to the NASA announcement, applicants had to be qualified in one or more of four career fields: aviation, medicine, science, or engineering. In aviation, experience flying high-performance aircraft was preferred, and special credit given those with test-pilot backgrounds. In medicine, general practice, internal medicine, and flight medicine were preferred. Scientists would be considered from all fields, but physicists and geologists were most highly regarded. All engineering specialties would be eligible, with extra credit given to aeronautical, nuclear, and electrical engineers. Advanced degrees were desirable but not mandatory, and for the American couple proof of U.S. citizenship was required. There was no minimum age, but there was a firm maximum of forty as of the date of application.

These criteria were not unlike those for earlier classes of astronauts, but the added ingredient of marital status made this selection process quite different. The restriction was considered unconstitutional by some, and a number of lawsuits were filed. Other critics pointed out that marriage can be a fragile arrangement and might not withstand the pressures of ten years of intense training before the flight actually departed for Mars. Nonetheless, NASA proceeded to evaluate husband-wife teams only, looking for pairs who embodied as many as possible of the disciplines required.

In this manner, NASA compiled a list of a dozen pairs of finalists. Eventually only one of these would fly. The third and fourth positions allocated to the United States were turned over to the Japanese, who were financing the design and construction of the Martian descent/ascent module. Japan had no difficulty in agreeing to the husband-wife arrangement.

A similar process took place in the Soviet Union, except that the Soviet Academy of Sciences decided to reserve for itself the right to pick all four of its crew members, two of whom would be Soviet citizens, while the other two would come from member nations of the European Space Agency. Canada, India, and Brazil were not eligible to propose crew members, although they were contributing hardware to the expedition.

When lists of candidates had been compiled and evaluated by the United States and Japan, NASA personnel met in Moscow with their Soviet counterparts and compared results. From this meeting the following agreement emerged as to which disciplines would be assigned to each side: The Soviet-ESA foursome would consist of one pilot, one flight surgeon, one nuclear engineer, and one physicist. The American-Japanese team would contribute one pilot, one internist, one electrical engineer, and one geologist. The landing party would consist of pilot commander, flight surgeon, electrical engineer, and geologist.

On October 31, 1994, sixteen individuals (eight forming the prime crew plus an equal number of backups) were introduced to the world at a press conference in Washington. The prime Soviet pair was Vladimir, a military test pilot and the mission commander, and his wife, Elena, an experienced flight surgeon. From the European Space Agency came nuclear engineer Henri and his wife, Marina, a physicist. They were both French. The Americans were Wesley, a geologist, and Heather, an electrical engineer. The Japanese couple were Sachi, a pilot and mechanical engineer, and his internist wife, Mariko. The selection of the sixteen was the result of intense secret negotiations between the Soviets and Americans. As a concession for having selected a Soviet mission commander, English was adopted as the crew's official language (almost a foregone conclusion anyway). But more important, Houston would be the location of the integrated training center and Mission Control.

All sixteen spent their first week on the job in hospitals, having their appendixes removed. Specialized training followed at those spots around the world where the Mars machinery was being built or where the various experts resided. Geology field trips were conducted in Iceland, Hawaii, Oregon, Mexico, and the Gobi Desert. Some training was done in pairs, quartets, or individually; on other occasions, all sixteen gathered together. The most intense training was done as a crew of eight. The prime crew spent several midwinter months as sole occupants of an isolated station on the Antarctic ice pack, a kind of shakedown cruise that revealed some hidden talents—and shortcomings. Mariko could somehow make their standard rations taste better than anyone else. But when it came time for Sachi to assume cooking duties, he balked, and it took considerable arm twisting by Vladimir to obtain his cooperation in those chores he considered beneath his status as a professional aviator and engineer. Fortunately Sachi looked up to Vladimir, a more experienced pilot, and eventually assumed his share of the housekeeping tasks. But Sachi seemed happier when pursuing individual studies, of which there were plenty.

In the Houston classrooms they all studied astronomy, rocket propulsion, space physics (with emphasis on solar flares), guidance and navigation, trajectories, rendezvous theory, and psychology. Each crew member had two basic responsibilities, and sometimes they collided. The first was to stay proficient in his or her professional field, and the second was to branch out and learn other disciplines that would be necessary for the safe conduct of their mission. Like it or not, they were becoming heavy-equipment operators. That was inevitable, given the fact that for twenty-two months the eight of them would be on their own in a cluster of large and complex machines on which their lives would depend. Mariko, accustomed to ferreting out human ailments, could not lose that skill, but to it she had to add diagnoses of a different sort: Was the beast slumbering peacefully, or was tank pressure rising ominously? What medication to prescribe? When to pull the circuit breakers on the heater circuits? Duties were further complicated by the fact that once in orbit around Mars, there would be only two in the Soviet round-trip module and two in its U.S. counterpart. Vladimir, Elena, Wesley, and Heather would depart for the surface, leaving Henri and Marina in charge of the Soviet module while Sachi and Mariko took care of things aboard the American craft. As the years of training went by and the crew gained experience, their original professional identities became blurred and they thought of themselves as simply space travelers rather than geologists or medical doctors. Except perhaps Vladimir. As commander and the one who would be doing the actual flying during descent and landing, his job seemed more an extrapolation of what he had spent his professional career doing. For the others it was more of a transition.

To Wesley their training seemed like painting the Brooklyn Bridge. They went over the spacecraft's design and the details of their mission from A to Z. The problem was, by the time they got to Z they had forgotten A and had to begin all over again. Fortunately there would be plenty of time—eleven months—while they were outward bound for Wesley to renew his knowledge of the hardware and to rehearse over and over again the tasks he would perform on the Martian surface. Although he didn't discuss it with anyone except his wife, Heather, Wesley felt that as the only geologist on board his contributions would be key, and that the others were supporting characters in the cast. Heather, ever the engineer, would reply, "You're only as good as your equipment, Wes." And Wesley had to admit that Heather was a quick study, that on their geology field trips she occasionally noticed something he had missed, some little detail that turned out to be vital in interpreting the rock formations they had seen.

Oddly enough, the French couple, Henri and Marina, spent more time in the Soviet Union than Vladimir. They had to become experts on the round-trip module manufactured there, while Vladimir had responsibilities spread over half a dozen countries. Henri, at home with complicated nuclear reactors, found the Soviet spacecraft a relatively simple design except for the closed environmental life-support system that recycled oxygen and water. Marina ended up as the crew CELSS expert, although she registered many mock complaints about it. "Not decent work for a physicist," she lamented, "all those dirty pipes. I will be the first space plumber." Henri and Marina visited often with Elena, who was honing her medical skills doing part-time work in the shock and trauma center of a large Moscow hospital. Elena shared with them the latest postcards from Vladimir as he gallivanted around the world.

One chore most of the crew disliked was their periodic meetings with the press. Mariko tried to clam up completely; if forced to speak she gave monosyllabic replies. Wesley tended to be too expansive, prating on at great length about the various rock types he expected to encounter. Geology interested the press not at all. What they wanted to know from Wesley was whether or not he expected to encounter life on Mars, and if so what kind, and what did he intend to do about it? Many of the questions were designed to probe their emotional state. Embarrassed by this form of attention (they dubbed these the "How do you feel about the meaning of life?" questions), the crew generally deflected their answers into more comfortable channels. "How do I feel about the possibility of not making it home alive? Well, with all our training and all the redundancy of our equipment, I don't think . . . well, for example, we have just figured out a new way to use less oxygen on the surface . . ." "I wish they would stick to the facts," Henri complained. There was no doubt that each of the eight had a long, if secret, list of disturbing possibilities, but as the years passed they became more confident in themselves, their teammates, and their machinery. Mars seemed a little closer somehow.

Except for Vladimir, who had orbited in the shuttle *Buran*, none of them had flown in space. Toward the end of their training each spent a month weightless in Earth orbit aboard a space station: Wesley, Heather, Sachi, and Mariko aboard

Freedom; Vladimir, Elena, Henri, and Marina aboard *Mir*. Half of them were nauseated for a couple of days, as predicted, but the glorious spectacle of the Earth soon restored their good spirits. Afterward they all agreed, somewhat somberly, that the black void between Earth and Mars would present an entirely different dimension of isolation, with no equivalent of the beautiful Earth out their window.

But if there were short spurts of excitement in their training, there were also long stretches of tedium. They spent countless hours in airplanes, between training sites, and even more time in unappealing motels. It was very helpful at these times to have husbands and wives traveling together, although those couples with children (the French and Soviet) had some extra worries on the home front. Fortunately all the children were old enough to attend school, and Elena and Vladimir's teenage son constantly pestered them to add him to the crew.

The bulk of their training was accomplished in simulators. For each flight module there was a replica in either Houston or Moscow. The cockpits of these simulators were virtually identical to those of the real machines, except that the simulated instruments were connected to banks of computers that could imitate normal and emergency conditions. All simulators were linked electronically to the Mission Control building in Houston, where a central staff could conduct practice sessions for various phases of the flight with crew members who might be as far away as Moscow or even Antarctica. For example, one dress rehearsal—the most complicated one—was a simulation of the events surrounding the Mars landing and subsequent exploration. For that, Sachi and Mariko were in the American module simulator in Houston, Henri and Marina in its counterpart in Moscow,

Vladimir and Elena in a duplicate of the Mars surface habitat in Antarctica, and Wesley and Heather in a replica of the Mars roving vehicle in the desert near Mojave, California. All were hooked by satellite communications to Mission Control, but with the addition of the nine-minute time delay they would have to contend with in the real case. Much was learned, especially in regard to communications and navigational links between the two orbiting elements and the two ground parties. It turned out that Mission Control was—compared with all previous manned flights—pretty much a helpless bystander under these conditions: the nine-minute delay eliminated them from any fast-breaking developments near Mars. Of course their experts were still helpful for long-term advice. In a foray outside their habitat to repair a leak, Elena suffered minor frostbite to her toes, the closest thing to a casualty during any of their training (a disappointment, perhaps, to the backup crew, none of whom was needed).

Finished spacecraft components began appearing on schedule early in 2003. On the Soviet side they were launched on Energiya rockets from Baikonur Cosmodrome to space station *Mir*. U.S. components went from Cape Canaveral aboard the Advanced Launch System to *Freedom*. Assembly began at both locations, initially by astronauts and cosmonauts who were specially trained in extravehicular activities and assembly techniques. As the various modules began to take shape, Mars crew members began to participate. Henri and Marina were launched aboard the shuttle *Buran* to *Mir*, where they spent several weeks checking out the Soviet round-trip module. Shortly thereafter Vladimir and Elena replaced them when the ascent/descent module arrived at *Mir*. Henri had to perform double duty during the assembly months, because he was also responsible for the installation and checkout of the surface nuclear power plant, which took place at *Freedom*. Henri was the first person to fly aboard both the Soviet and U.S. shuttles. At *Freedom* he joined Sachi and Mariko, who were working on the U.S. round-trip module. Wesley and Heather followed to make sure the Rover, as the surface vehicle was called, was properly installed.

After a busy year of assembly and checkout of the two clusters of machines, each mounted on its gigantic aeroshell, the tedious but dangerous process of loading supplies and fuel began. Mariko and Marina were put in charge of the inventory and control of consumable items, of which there were more than one hundred thousand in each round-trip module. The library of data disks, printed manuals, checklists, and cassettes totaled more than six thousand items. Food for nearly sixteen thousand meals had to be organized for a minimum of confusion in flight, since unpacking and repacking in weightlessness could result in a storm of projectiles flying about the cabin. Washcloths, toilet paper, vitamin pills, adhesive tape: were small items like these community or individual property? Vladimir and Elena, both members in good standing of the Communist Party, argued strongly in favor of individual ownership, to the amusement of Wesley and Heather.

More serious was the matter of loading highly combustible oxygen and hydrogen. Oxygen liquefies at 293 degrees below zero, hydrogen at 423 below. Transferring fluids at these temperatures from *Mir* and *Freedom* into the storage tanks was a delicate and dangerous operation; care had to be taken that there were no leaks or boil-offs and that all tanks were filled to maximum capacity. Filling the water tanks was equally vital, but a lot easier.

In mid-May 2004, a final training exercise took place, with four crew members at *Mir* and four at *Freedom*. Nearly ten years had passed since the eight had been selected. As individuals and as a team they had made great strides, and they were able to handle nearly all the simulated emergencies with which Mission Control confronted them. Despite minor flare-ups between Vladimir and Sachi, their lives together had been amazingly harmonious. Yet each of them knew that the flight could not be merely the sum of the parts of their training. The next twenty-two months would be totally different, and each crew member felt some apprehension. There was so much to know, to retain. There were so many unknowns, not the least of which was how he or she would react in times of stress or crisis. They were about to find out.

CHAPTER

XXII

OUTWARD BOUND

AFTER MUCH debate the crew named their round-trip craft *Sigma,* the eighteenth letter of the Greek alphabet and the symbol mathematicians use to signify summation or totality. They liked the symbolism because their great pile of machinery *was* all that they would possess for nearly two years. It would be the summation of their existence. Also it was a neutral concept that should offend no one's religion or national traditions. For the descent/ascent module, flying in and out of the Martian atmosphere, they wanted the connotation of wings, so they settled on *Cygnus,* the Latin name of the constellation in the Milky Way that the ancients thought looked like a swan. Like *Sigma, Cygnus* was inoffensive. Unfortunately, during simulations they discovered that the two words sounded almost identical over the radio, and much confusion resulted when Mission Control tried to call one or the other. *Cygnus* was changed to *Condor,* the gigantic mountain flyer of the Andes. Its South American origin pleased the Brazilians, who had built substantial parts of the descent module and who were miffed that they were not represented among the crew. *Sigma* really would come into being only after they had departed Earth and its Soviet and U.S. halves were joined and docked. Until then the Soviet craft, which was slightly heavier and would be the first to set course toward Mars, was called *Sigma 1*; the U.S. counterpart was called *Sigma 2*.

On June 3, 2004, *Sigma 1* separated from *Mir* and ignited its own engines for the first time. After a thirteen-minute burn it had established itself on a trajectory to Venus, five and a half months away. Three hours later *Sigma 2* undocked from *Freedom* and followed, as planned. By that time *Sigma 1* was already seventy-five thousand miles from Earth, and it would take *Sigma 2* nearly ten days to catch up at a leisurely pace. The two crews soon noticed a change. In Earth orbit, the view had been somewhat like the terrain going by beneath a jetliner. True, those were oceans, not lakes; continents, not counties—but still, the ground slipped by and there was a feeling of speed. After leaving Earth orbit, although they were going faster than ever, the feeling of speed diminished and then stopped altogether. They could see the Earth getting smaller, rapidly at first and then barely perceptibly. It was an eerie, sobering sight, their homeland receding. They had a heightened awareness of distance, and of time, but somehow not of speed.

After the excitement died down and they had packed away their cameras, the crew settled into their new home. Heather remembered how tiny the interior of *Sigma 2* had seemed the first time she sat in it at the factory. And then in orbit, how much larger, and how different it had seemed. Once, she came back into the wardroom from the bathroom and, confused, turned the wrong way and flew right into the wall. In weightlessness a lifetime of rules no longer made any sense. One didn't have to walk on the floor and look *up* at the ceiling. One could plant one's feet wherever one chose and regard *any* direction as up. That was why, at first, Heather found the lettering on one instrument panel "upside down." She had instinctively approached the panel via the shortest route, which happened to be from what on Earth would be the ceiling. On Earth, gravity pins us to the floor and causes us to waste interior space. We never use all that good room above the windows up next to the chandelier. Heather was fascinated by weightlessness. As the smallest and probably weakest crew member, she found weightlessness a great equalizer. She could move around *Sigma 2*, even with bulky equipment, as well as anyone else, perhaps a little bit better. She found lovely nooks and crannies that fit no one else. During movies, for example, she settled her five-foot frame into a spot opposite the screen, with her back against the main circuit-breaker panel and her feet wedged up against the sextant bracket. This became "Heather's corner" and no one else invaded it. These things happened naturally aboard *Sigma 2* as the crew settled in. There wasn't much talk about it. Life aboard *Sigma 1* was more organized. Vladimir believed in discussion, consensus, and group action. Wardroom spaces were assigned on a functional basis, Vladimir near the communications equipment, Marina next to the CELSS control panel, and so on.

As *Sigma 2* slowly overtook *Sigma 1*, the tension built. If the docking was not successful, they would probably survive, but no landing would be possible, for the equipment attached to both halves was essential for operating on the surface of Mars. Two midcourse corrections were required as *Sigma 2* closed in, with Sachi firing the thrusters while Heather, glued to the radar, announced the steadily decreasing distance between the two. Then, when they were roughly a hundred feet apart, Vladimir took control in *Sigma 1* and guided the two ends of the docking tunnel together. Capture latches snapped with an audible click, and *Sigma 1* and *Sigma 2* were merged into one mother ship. "Mission Control," Vladimir reported in a raspy, Slavic accent, "*Sigma* here. It's okay, the docking, it's good."

Work aboard *Sigma* was organized by shifts. Panel clocks and wristwatches were set on Greenwich mean time. The first shift was from 10:00 A.M. to 6:00 P.M. Greenwich, the second from 6:00 P.M. to 2:00 A.M., and the third from 2:00 A.M. to 10:00 A.M. The majority of the work was scheduled for the first shift, although the second could also be fairly busy. The third shift was for sleeping only unless something like a timed hardware test intervened. The idea was to keep the lives of the supporting engineers back on Earth as normal as possible. Setting the key shift for 10:00 A.M. to 6:00 P.M. Greenwich time was a compromise: with Greenwich lying between Houston and Moscow, the Houstonians wouldn't have to arrive at work too early or their Moscow counterparts too late. The crew members didn't care one way or the other, because their normal *zeitgebers* were missing. Out their windows the Sun shone twenty-four hours a day, while at the very same time by shading their eyes and looking away from the Sun they could see stars. Within a few days they could adjust with equal efficiency to any schedule

Mission Control desired. Time was just little numbers on a watch, thought Marina, although she liked the idea of not having to report to work until 10:00 A.M. It seemed civilized somehow, almost decadent. But at ten o'clock sharp, Vladimir expected everyone to gather in the wardroom of *Sigma 1* for the daily meeting. The Soviet designers had selected a grayish green color for the wardroom, while over in *Sigma* 2 their American comrades had chosen lemon and sand. The two facilities, each the heart of its cluster of modules, were always referred to as the Green Room and the Yellow Room. The Green Room had a better layout for roundtable discussions, the Yellow Room for audiovisual presentations. The 10:00 A.M. Green Room meeting was the only one at which attendance by all eight was mandatory. Vladimir presided, discussing the status of systems and the schedule for the day. Sometimes he included a lecture on one of the subsystems, delivered by the crew member responsible for it. Everyone hated it when Vladimir's turn came because he invariably talked too long, sometimes for two or three hours. Then Wesley would interrupt with a "question for the commissar." Vladimir got the point but kept on talking anyway. On Sundays the meeting was held as usual, but after a brief systems check they were dismissed. Mission Control tried hard not to bother them on Sundays with instructions or requests.

From time to time their daily routine was broken by a "sim," or simulation. Sometimes these were scheduled, and sometimes Mission Control sprang them without warning. A Klaxon would sound, and the crew would rush to the wardrooms, where warning light panels and the diagnostic computers were located. A flashing SIM light on the panel was the crew's notification that the emergency was not real, but they treated sims with great seriousness. It was a matter of professional pride for them to solve each imaginary problem as quickly and thoroughly as if it were the real thing. Their adrenaline actually flowed and they could feel their pulses quicken and their breathing grow shallow. Their egos, if not their lives, were at stake. An unscheduled sim was a major event, something to be discussed over and over at the dinner table.

The sims were possible because of an ingenious software program built into their computers. It allowed the spacecraft instruments to be disconnected electronically from their normal sensors and to be wired instead into the sim box, a software center that contained instructions relayed by radio from Mission Control. Mission Control was fiendishly clever in inventing subtle malfunctions, ailments that infected more than one subsystem in a seemingly unrelated pattern. "A drop in number seven O_2 tank pressure," reported Henri, "and also number two H_2 quantity shows zero. Could we have had an explosion, or maybe a meteorite strike, that has taken out both tanks?" "No way," replied Elena. "If that happened you'd see increased current flow in the heater circuits. They're all normal. Those two tanks share the same signal-conditioning unit. It must be faulty. Try switching Essential Power A-4 from Primary to Secondary. See! That fixed it, huh?" Elena was a whiz at sims, perhaps because of her training as an emergency-room physician. Vladimir was proud of his wife, but it galled him nonetheless. Despite years of exposure to flight hardware, and several actual in-flight emergencies, Vladimir felt dull and slow compared with Elena. Even worse, she loved to relive the sims, savoring each clue and her analysis of it. Then Vladimir would resort to rapid-fire Russian. The others had an idea what he said, but it didn't seem to faze Elena. Vladimir had some misgivings about being more vulnerable to actual disasters during sims because of the interruption of information from the many subsystem sensors. But he had to admit the exposure was brief and the training valuable, not to mention an unexpected by-product: the sims were a great antidote to complacency and boredom. More than any other device they maintained crew motivation, coordination, and morale.

Milestones were also important morale builders. Their first celebration was on July 9, midway between the American Independence Day (July 4) and the French Bastille Day (July 14). In honor of the joint occasion, Mission Control agreed to a Sunday schedule and opened up the radio network to relatives and friends of the crew. Familiar voices poured into *Sigma* all day. It now took radio signals, traveling at the speed of light, about twenty seconds to reach them from Earth, so normal conversations were not possible. One could ask a question and wait forty seconds for an answer, but instead one usually got the tail end of a conversation begun before the query was received. But despite the confusion it was fun. Instead of four eating in the Green Room and four in the Yellow, all eight gathered in *Sigma 2*'s wardroom to enjoy one of Mariko's extra special meals—octopus and vegetable noodles. Granted the food was frozen, not fresh, but Mariko's spices made it seem like four-star fare. At the conclusion of dinner Mariko produced another treasure: eight disposable hypodermic syringes filled with an amber fluid. She had raided the infirmary and put a couple of ounces of precious medicinal brandy into each syringe, cutting off the needle at its base. As she demonstrated, one could squirt a tiny stream of brandy into the mouth without losing a drop. No one had tasted alcohol for over a month, and its effect was immediate and friendly. Vladimir, ever on duty, gave most of his to Elena, but the others savored theirs to the last drop, as conversation in the Yellow Room went on and on. For the first time since leaving Earth, they were relaxed enough to discuss matters other than their flight, and they reminisced about earlier, less complicated days.

Life on board *Sigma* settled into a routine that was confining but well above the threshold of their tolerance. Originally *Sigma 1* and *Sigma 2* had each been designed with four tiny sleep chambers, but one of the few changes the crew had insisted on was to modify these into double compartments. They were still tiny by Earth standards, but each couple's work schedules were such that, except for when they were asleep, usually only one person was there at a time, and then the compartment seemed quite commodious. Except for a few restrictions—no fire hazards, for instance—each couple had had complete control over their compartment's furnishing, making them the one part of the spacecraft that had a warm and personal touch. Sure, there were a couple of vines growing out in the wardrooms, but here was a little piece of home, with mementos and decorations that meant a lot to each of them. Like small children, they had "their room." Their marriages, sound to begin with, continued to be so, although there were the inevitable squabbles. The couples discovered they had one disagreement in common—the matter of sleeping "head toward" or "head away." Two ventilation fans pumped air into one end of each sleep compartment. A double sleeping bag (or "containment assembly," as it was called on their computerized stowage list) was stretched along the length of the compartment, and it could be rigged with the head at either end. Thus one could sleep "head toward" the flow of air or "head away." The arguments were endless. Most agreed that "head away" was warmer except for the feet, but the airflow then blew up the nose and was thought to be bad for the sinuses. Once Marina even got a nosebleed after a night in that position. Yet "head toward" was noisier, and the air tended to penetrate the sleeping bag, not to mention the distraction of hair blowing across the face. Henri and Marina had the worst fights about it, and Henri, quite serious, threatened to rip the bag down the middle. Finally, family harmony was restored when he assembled a makeshift cardboard pipe ("the elephant trunk") that diverted part of the airflow.

Each room was slightly different, in shape and in noise level, depending on what machinery was contiguous to it. Some had pipes that gurgled contentedly through the night, while Wesley and Heather had something they called the "thumper," apparently a solenoid valve that cycled on and off at irregular intervals. Like the crew on a nuclear submarine, the *Sigma* crew grew accustomed to the presence of some noises and the absence of others. Animated conversations heard but not understood were universally disliked. Raised voices upset the serenity of their voyage and implied trouble. Unconsciously the crew adapted to this perception and generally spoke to each other in hushed tones—which made it that much worse when someone did shout. Then people would converge, even tumbling sleepy-eyed from their bedrooms, to find out what was going on. Vladimir was the worst shouter by far.

Late one evening in mid-October the Klaxon sounded in *Sigma 2* and they all gathered in the Yellow Room, expecting to see the SIM light flashing. Instead they found the main monitor screen blinking WASTE P. The differential pressure in a waste line was beyond its normal limits. Heather, first on the scene, called up the computer display for CELSS and located the sensor that was giving the abnormal reading. It was in the urine line and indicated a partial blockage. The occupants of *Sigma 2* were not surprised. For some days they had noticed that their toilet seemed to be ailing. The bathroom was a place to avoid. The toilet gave off a foul odor, and its flushing mechanism, controlled by a strong blast of air, was slowing down. The airflow was as vigorous as ever, but it took longer for the trapped blob of urine to disappear through the exit screen. The computer referred them to page 43 of Manual VI, but that was just a schematic drawing of the urine-recycling subsystem and gave no clue as to what to do about this particular problem. Mission Control told them to "stand by," which they did for more than an hour while arguing among themselves. Wesley suggested running some comparative tests with *Sigma 1*'s system, which was functioning normally, but Elena pointed out that the two were far from identical and the results might be deceiving. Finally Mission Control suggested that for twenty-four hours they all use *Sigma 1*'s bathroom and pour water into *Sigma 2*'s urine collector at prescribed intervals. They agreed and went back to bed feeling uneasy. "I don't know," Wesley said to Heather. "Four months down, eighteen to go, and already one john has crapped out."

The Earth was so far away now that it looked just slightly brighter than a first-magnitude star, just a bluish white pinpoint of light. Venus, on the other hand, was *bright,* a small yellow half circle, looking like a lemon that had been cut in two. It was more and more in their minds and in their conversation. Ever the scientist, Wesley was the most intrigued by Venus and was looking forward to using an array of instruments to measure and photograph it from a distance of four thousand miles, their closest point of approach as they swung around Venus and used its gravity field to speed them on their way to Mars. Wesley knew that Venus's thick atmosphere would prevent even a glimpse of its surface, but the clouds themselves were of great scientific interest. Composed mostly of carbon dioxide, but with a little bit of nitrogen, oxygen, and water vapor mixed in, light and dark cloud tops swirl and eddy in a motion that takes them once around the planet every four days.

Unlike Wesley, Vladimir did not like Venus. To him it was a worry, a problem that centered on temperature control. Venus was hot as hell. The heat coming from it, mostly reflected sunlight, would cause problems, he was sure. The safe thing to do was to point one of the aeroshells directly at it, hiding the delicate machinery in the cool shadows. But there were problems with that too. It meant no photographs or visual observations because the aeroshell would block the view. Also, having the aeroshell in that orientation would prevent communicating with Mission Control, where hundreds of scientists were gathered to learn as much as they could about this strange planet whose dense atmosphere produces an intense greenhouse effect. Before the flight Vladimir had discussed his concerns at great length with Mission Control, whose experts were divided on the best course of action. Finally they had decided to approach aeroshell first, and then to rotate to best photographic position for about twenty minutes. Now Vladimir assigned himself in *Sigma 1* and Sachi in *Sigma 2* to be temperature monitors. They would spend those twenty minutes glued to their computers, checking each temperature-critical subsystem in turn. Sachi scowled when Vladimir informed him of these duties. Sachi had done a lot of scowling lately, Vladimir thought. Except for that night with the brandy, when he had been laughing almost hysterically.

For the first time since leaving Earth, they began to feel that they were speeding through the sky, not just hanging weightless and motionless in their cozy cocoon. Venus was a large disk now, half in sunshine, half in shadow, and they flocked to the windows to watch it grow. Even Vladimir had to admit it was beautiful and ever-changing, a creamy swirl of clouds bubbling with hidden energy. November 20 was the day of closest approach. The evening before, their temperature gauges told them it was time, and reluctantly they swung the heavy craft around until *Sigma 1*'s aeroshell faced the planet, blocking their view. It also blocked their radio antenna, and they lost contact with Mission Control. Waiting in silence was an eerie, blind feeling; they sensed the giant outside but were unable to confront it. Temperature readings dropped reassuringly, but Vladimir thought he could hear crackling noises coming from the aeroshell's expansion joints. No one else heard them. Finally, exactly on schedule, they maneuvered to place the wardroom windows in viewing position. At first they were confused by the sight, as their eyes tried to adjust to contrasting conditions. *Sigma* was in Venus's shadow, and the clouds looked dark and sullen now, but a thin arc of piercing sunlight cut diagonally across their field of view. This halo came from behind Venus, where the Sun's rays struck the clouds and, diffused somewhat, cascaded around the planet's rim. Yesterday's flat disk was gone, replaced by a chunk of three-dimensional sphere whose bulging belly seemed to be trying to push its way into their cabin. It was an ominous presence, no longer a cheery lemon pie in the sky.

Less than a minute after Venus swung into view, inquiring voices crackled on the radio. As best they could, the crew described what they were seeing. Telemetry from various spacecraft instruments would augment their verbal reports, but even more important were the data Wesley was recording with cameras and spectrometers. Every wavelength was being measured, every viewing angle examined. "Hey, this is worth the price of the trip," Wesley exulted. "Isn't that what Armstrong said when he got to the Moon? I just wonder why those guys didn't describe it better."

Beyond Venus, life aboard *Sigma* quickly became routine again. Their excitement ebbed as the yellow planet shrank and once again became an ordinary object in their windows. Their mood was quiet and serious. Levity seemed out of place. Even their Christmas party failed to raise their spirits. In fact, Christmas was the worst time they had known so far, a time for remembering Christmases past—children, old friends, presents, fireplaces, trees, dogs. These memories were unsettling, and although Mission Control did its best to cheer them by beaming up the latest TV comedy programs, nothing seemed to work. January was a bad time. In one direction they could see Venus and the Earth, nearly superimposed; in the opposite, Mars. None seemed significant. They felt no motion, no progress. They just hung there, bored and listless. With a felt-tipped pen,

Wesley drew a crude calendar on the bedroom wall and marked off each day. He and Heather got more reassurance from staring at it than they did from watching their television screen.

Vladimir organized games and competitions. For chess he worked out an elaborate handicapping system that determined to what extent each player could call upon expert advice from the on-board chess software or by consulting grand masters on Earth. With a little help from sixty-eight-year-old Boris Spassky, Elena won the first tournament, much to Vladimir's dismay, for he considered himself far more skillful than the others. For the next tournament he changed the rules somewhat, but Mariko, who had never played the game on Earth, won. Vladimir moved on to backgammon.

Getting everyone to exercise two hours a day on the treadmill and stationary bicycle became more difficult. It was such drudgery, although they did admit they felt better after it was over. Some studied while they exercised by selecting training tapes for the TV screen mounted next to the exercise machines. Others listened to music through earphones. Sachi simply stared at the wall. One minor but annoying problem was disposing of sweat. In weightlessness big globs of it formed on their faces and got into their eyes. If they shook their heads, spheres as large as golf balls flew off and splattered against the walls, instrument panels, and equipment lockers. To the two physicians, Elena and Mariko, this was unsatisfactory from a hygienic perspective. Heather, the electrical engineer, worried that the errant moisture might find its way into delicate equipment and cause a short circuit. They tried wrapping towels around their heads, but that was too hot and caused even more sweating. They jury-rigged a fan overhead and that helped somewhat by blowing droplets onto their jackets. But in the end they gave up and simply wiped their heads with a towel once every minute or so for two hours. Having to do that added to their irritation.

In addition to the treadmill and bicycle, another device aided their physical well-being. Nicknamed penguin suits, these coveralls were custom-tailored to each individual with heavy bands of elastic looping from legs and crotch up around the shoulders and upper arms. When zipped tightly into one, a crew member had to fight against the suit to remain upright and to move the upper torso. The suit put a compressive load on the spine and its supporting muscles and exercised the muscles of the legs, shoulders, and arms. It was thought that the penguin suit would slow down the processes of bone demineralization and muscle atrophy. Unfortunately the suits tended to be hot and uncomfortable. Crew members were supposed to wear their suits for the six hours of each duty day that they were not on the treadmill or bicycle, but frequently this period was cut short, and they used the slightest excuse to slip into loose-fitting coveralls. Marina liked to zip around in her underwear.

The situation with *Sigma 2*'s plumbing probably had something to do with their attitude toward exercise and the penguin suits. The obstruction in the toilet's exit pipe had gotten worse. Pouring water into it had not helped, nor had the various combinations of medicines they had tried flushing down it on the assumption that the blockage might be biological in nature. Whatever was causing it, the blockage was too far away from the cabin to reach by disassembly:

the shower wastewater also refused to drain, which meant the main line was clogged at a point downstream of the juncture of the two systems. No one had thought to include any drain cleaner, suction cup, or plumber's snake on the stowage list. "Stupid CELSS," muttered Sachi. "Why did they route those two lines together, anyway?" "They had a reason," replied Marina. "They thought that the shower water would dilute the urine and make it easier for the filters to handle one fluid rather than two." Sachi was not convinced: "Stupid CELSS!"

The shower and the toilet in *Sigma 1* now had to serve all eight. Vladimir did not want to overload it, and he could not control their urine output, so he decreed that the interval between showers be stretched from one week to two. It was shortly after this decision that the grumbling about sweaty exercise grew noticeably louder.

Starting on March 3, 2005, the crew underwent their ninth-month physical examinations. This schedule had been selected after much debate over the years, with Soviet and American physicians poring over *Mir* and *Freedom* data and conducting lengthy medical conferences—usually at Key West during the winter and in the summer at Alupka on the Black Sea. Nine months was seen as the ideal compromise. The doctors wanted to wait as long as possible for ailments (caused by weightlessness) to appear, yet they wanted to know as soon as possible if anything was wrong, so they would have enough time for treatment before the Mars landing, which was scheduled for May. Each physical took a whole day and was as comprehensive as possible, given that the on-board equipment fell far short of what a large hospital or medical laboratory would have. They did not have an X-ray machine, but they could analyze blood and urine samples. Mariko, an experienced internist, conducted the examination of all seven; hers was performed by Elena, who was more accustomed to treating trauma or illness than diagnosing it.

Examination results were similar. They all had abnormal red blood cells called echinocytes. For some unknown reason, weightlessness causes these star-shaped cells to be produced in the bone marrow. When the crew returned to Earth, the echinocytes would slowly disappear. What effect Mars's reduced gravity would have on them, no one knew.

Second, all showed increased urinary concentrations of nitrogen, calcium, and hydroxyproline. The nitrogen meant that despite their exercise they were losing muscle mass (measurements of calf circumference verified this, and they could confirm it just by looking at their skinny legs). The presence of calcium and the amino acid hydroxyproline indicated a more serious problem—the loss of bone. Their skeletons were slightly less dense now, especially the bones of the lower back, legs, and feet. Too much calcium in the urine could also lead to kidney stones, a painful and potentially life-threatening ailment. If worse came to worst Elena would have to try surgery to remove a stone, a complicated operation that she had watched but never performed. If Elena came down with a stone, Mariko would have to give it a try. On Earth stones can be dissolved by ultrasonic vibration, but they had no equipment for that.

Third, their cardiovascular systems had deconditioned. This fact was determined by the lower-body negative pressure test. This involved a barrellike device that enclosed a crew member's lower body and was sealed by a rubber skirt at the waist. The air inside the barrel was vented, and the vacuum sucked blood from the head and torso down into the legs, just as gravity tried to do on Earth. The intensity and the duration of the vacuum that one could endure without fainting was a measure of cardiovascular condition. Elena did better than the rest on this test ("because she is a little round sausage," according to Vladimir). Tall, lanky Henri fared the worst. Since he was not scheduled to visit the surface of Mars, his condition was not scrutinized as closely as that of the landing party.

Fourth and last, the eyes of several crew members exhibited "submarine syndrome," a nearsightedness that comes from months of looking only at objects within close range—in this case within the confines of the spacecraft. The condition had been anticipated, and the medical equipment on board included an optical training device. When held up to the eyes, it presented targets that cycled from a focal distance of twenty feet out to infinity and back. Vladimir and Elena were now scheduled for daily use of this equipment. In addition they would have two weeks in Mars orbit prior to landing, a period during which their observation of the Martian surface (at an optical distance of infinity) should easily restore their vision to its preflight norms.

These results were all expected, and they were within predicted limits, although the doctors back on Earth (after a quick meeting in Key West) urged them to increase their daily exercise routine to two and one-half hours. In addition there were various minor complaints. Dry skin with chafed, irritated patches was prevalent. They had already used their few bottles of moisturizing cream, which had been regarded as an unnecessary frill. The atmosphere inside both habitat modules, but especially *Sigma 1*, was very dry, and several of them had had nosebleeds. There were complaints about the food—its blandness and the repetitive nature of their menu, the lack of fresh fruit and vegetables. "Oh, for a lemon, an honest-to-God lemon." Flatulence was a common complaint because of the gas bubbles that somehow found their way into *Sigma 2*'s drinking water. Some made it a habit to go all the way through the tunnel into *Sigma 1* just to get a drink of water.

Also considered part of the physical exam was an evaluation of the radiation dose each crew member had received. Dosimeters had been placed in a dozen locations around *Sigma 1* and *Sigma 2*, including one in each sleep chamber. In addition, crew members had a tiny dosimeter built into their wristwatch bands. A computer program added up the readings of each instrument, estimated where each crew member spent his or her time, and calculated individual doses. There were variations, depending primarily on how well shielded (by water and fuel tanks) each sleep chamber was, but overall the readings were lower than expected, and the crew rejoiced in that good news. The idea of radiation was somehow sinister, the realization that a lethal dose could pierce their bodies without pain or even awareness. They expected low dosimeter readings, but they were nonetheless pleased to have a computer printout so attesting.

The crew was pleased with the results of their physicals, and after a couple of days of chattering about how well they were doing, morale picked up. With Mars less than two months away, they saw it now as a plump red beacon, growing ever larger and more inviting. All in all, exam week had been a great success, although Wesley had to admit that—after he had been thoroughly, almost roughly, examined for testicular tumors—he looked at gentle Mariko in a slightly different light. Not better, not worse—just different.

On April 1, Heather broke her collarbone. She had come sailing out of her sleep compartment, traveling fast as usual, and somehow misjudged her direction as she rotated into the Yellow Room. To correct herself, she grabbed the sextant bracket with her left hand and cartwheeled face first into the main circuit-breaker panel. Her right shoulder slammed into it. The pain was slight at first, but within minutes grew worse, especially when she moved her right arm. Mariko took one look at her sagging right shoulder and clucked, "Ah so, Heather . . . collarbone, dear." Heather got a shot of morphine, and Mariko aligned the bone as best she could simply by feel. Then she strapped Heather's arm across her chest. That was all anyone could do.

Heather's physical pain soon disappeared, but it was replaced by a feeling of despair as she thought about what lay ahead. They were due to arrive at Mars on May 9, and expected to land around May 23. Seven weeks left! How long to heal a broken collarbone? "Oh, usually around six weeks, dear." What Mariko did not say was that even if the break healed perfectly, it would be susceptible to further injury for long after the six weeks. Furthermore, her right arm and shoulder would be weak, and Heather was not particularly strong to begin with. The others had nicknamed her "the Flea" because of her tiny stature and quick movements. Now the Flea was within weeks of the Martian surface, where she would have to operate inside a bulky pressure suit. She would need the full strength of both hands to manipulate the heavy equipment that she and Wesley were scheduled to use there. Was she up to it?

The matter was debated at length, both on board *Sigma* and back on Earth. The original plan called for Vladimir and Elena to remain at base camp while Wesley and Heather made excursions in the Rover. All their training had been based upon this division of labor, with the result that neither Vladimir nor Elena was capable of performing Heather's duties. As the chief decision maker, Vladimir felt uncomfortable going off on a long geologic foray in the Rover. He wanted to stay in camp, in radio contact with both the Rover and the orbiting *Sigma,* because he felt that was the central location from which he could best direct operations. That meant that if Heather was not capable of accompanying Wesley, Elena would have to go. Heather was a brilliant electrical engineer who knew the Rover inside out. Wesley was a field geologist and regarded the Rover as his truck, but he focused on science, not machinery. Elena was a physician and a clever troubleshooter, but she knew next to nothing about the Rover or the equipment it carried. Furthermore, the Rover was stowed in its own sealed compartment and could not be operated or even examined until after they landed. They could not wait for Heather if she required a prolonged recovery; they had no consumables to spare, and besides, the alignment of Earth and Mars dictated their departure date, which could not be delayed for more than a couple of days. As a precaution, Elena dug out all the Rover manuals, and she and Heather spent most of every day poring over them.

Their trajectory had been amazingly precise approaching Venus, but after swinging around it they had begun to stray from their desired path, probably because of miscalculations of Venus's gravitational field. A course correction was required using their main rocket engines instead of the small vernier rockets they had used near Venus. The problem was that in order to use these big engines, *Sigma 1* and *Sigma 2* had to undock and make separate firings. This was a tricky operation, requiring the closest coordination between Vladimir in *Sigma 1* and Sachi in *Sigma 2*. It was not something they wanted to do twice, so the trick was to pick the optimum point on their trajectory at which to perform the maneuver.

As is so often the case in spaceflight, the decision became a trade-off between uncertainty and efficiency. The sooner they made the engine burn, the shorter the distance they would have strayed off course and the less fuel required to make the correction. But the longer they waited, the more tracking data would be accumulated, which meant they would know more precisely what correction was required, making a second maneuver later less likely. Eventually Mission Control decided to sacrifice some efficiency for added certainty; they delayed the maneuver until they were fairly close to Mars. They wasted some fuel by waiting, but they felt they gained more by making sure the course correction was precise.

When the moment arrived, Vladimir undocked *Sigma 1* and moved about half a mile away. With a simultaneous countdown, the two craft ignited and shut down their motors within seconds of each other. Nonetheless they drifted some twenty miles apart in the process, and it took Vladimir the better part of a painstaking day to slowly close the gap and dock again. Then they relaxed once again.

One lazy afternoon in early May, within a week of the scheduled arrival at Mars, the Klaxon sounded. The SIM light stayed off, but the screen blinked SOLAR EVENT. This was not unexpected because for several days now, Mission

Control had included in its daily report the fact that unusual solar activity had been noted. A quick glance at their dosimeters showed only a slight increase, but apparently it had been enough to trigger their computer-activated alarm. The computer program analyzed dosimeter trends; an abrupt rise, even a small one, was cause for alarm. Apparently a solar flare was just beginning. Thirty years before, the primitive American space station *Skylab* had included a small observatory for measuring the Sun's activity. With powerful telescopes *Skylab* astronauts could witness the swirls and eddies in the Sun's corona, and they recorded storms several hundred times as large as the entire Earth. To save weight, *Sigma* had no such instrumentation, a fact that physicist Marina regretted now. Their telescope was designed to pick out detail on the surface of Mars, not to look into the blinding light of the Sun. The crew's only recourse was to retreat to their storm cellar, the tiny Earth-capture vehicle they had dubbed *Peanut*. *Peanut* was well shielded, being located underneath *Sigma 2*'s habitat module and supply tanks. With the aeroshell of *Sigma 1* pointed at the Sun, the high-energy protons streaming out from the solar flare would have to penetrate all of *Sigma 1* plus most of *Sigma 2* to reach *Peanut*. They were safe as long as they stayed put. It took solar particles eleven minutes to travel from the Sun directly to *Sigma*. It took a minimum of eighteen minutes for word of the flare to reach them from Earth (eight minutes for the Earth to find out, plus another ten for the warning to reach *Sigma*). Sure enough, about seven minutes after the Klaxon went off, Mission Control—trying not to sound worried—came on the air, describing what they already suspected. It appeared to be the preliminary phase of a powerful flare, definitely the largest so far in 2005. By the time notification arrived, they were already crowded into their storm cellar. *Peanut* was designed to hold all eight of them, but only for a matter of hours, just long enough for them to penetrate the Earth's atmosphere on return. It was very crowded, hot, and not very well lighted. All they could do was wait, with occasional quick forays to the bathroom or galley. Sometimes flares lasted several days. As the time dragged by, it became apparent that Mission Control had no idea how long the radiation might continue. But after about fifteen hours the flare's protuberances abruptly collapsed and disappeared into the Sun's interior. Shortly thereafter it was deemed safe to emerge, and dosimeter readings were collected. Pointed at the Sun, *Sigma 1* was hit hardest, with average recordings of 240 rem. *Sigma 2* had been penetrated by 55 rem, and each crew member had absorbed approximately 14 rem. This was almost as much radiation as they had accumulated in the previous eleven months, but still far below the safety threshold. The only other evidence that anything unusual had occurred was that one of *Sigma 1*'s electronic clocks had been knocked out of synchronization. The crew went back about their business much relieved. Most were jubilant that another test had been passed so easily, but Heather remained glum, and Sachi, as usual, seemed to suffer from the time spent in cramped quarters.

"Potatoes!" exclaimed Marina, turning away from the eyepiece of the telescope. "Yes," agreed Henri, "potatoes, baked . . . with sour cream maybe, and chives also." "No, silly—Phobos and Deimos. They really do look like blackened potatoes, like ones that fell into the fire." The telescope was now the most popular spot aboard *Sigma,* and the crew lined up to get a peek at Mars. After years of study they were mesmerized by the sight. It was exactly as it looked in the photographs, yet at the same time it was so different. Vladimir felt the experience was like switching from an old-style television set to high-definition TV, but the others hooted at that comparison. This was much more spectacular than any television image, they all agreed. They could easily pick out the largest features through the telescope and then with their naked eyes. The great crack known as Valles Marineris was particularly impressive as shadows highlighted its serpentine passage. The view Marina enjoyed the most was when she caught Phobos, the larger of the two moons, moving across the pockmarked face of Mars, a tiny black clipper ship passing in slow and stately review.

Suddenly the days of lazy looking ended. It was time to get busy, to separate the vehicles once again, so that each could aerobrake individually into orbit. They were glad now that they had had to make that large course correction. Because of it they felt a lot more comfortable with the idea of undocking, decelerating, and then finding one another again. When the time came, they separated by several miles and took a good look at the red sphere that filled their windows. It would be invisible to them as they plunged into its atmosphere, aeroshell first. Waiting was difficult, even more than it had been during the hours Venus was blocked by their aeroshell. This was different. This was their destination, not just a way station; and in addition, the maneuver they were about to perform entailed a degree of difficulty, a level of precision, exceeding anything they had done so far. Both vehicles needed to reach a stable orbit in order for the landing to proceed. It was true that *Sigma 2* alone contained *Condor,* the landing craft, but the supply ship carried by *Sigma 1* was needed to sustain the team on the surface. According to the mission planners, the visit wasn't sufficiently safe otherwise. Just to make sure that the crew of *Sigma 2* didn't get any ideas about making a brief landing should something happen to *Sigma 1,* the crew's training prevented it. Only Vladimir and Elena, aboard *Sigma 1,* were trained to fly *Condor*; Wesley and Heather were mere passengers during descent and ascent. No, it wouldn't do to have an international expedition result in one superpower succeeding and the other failing. The planners had made sure that both *Sigma 1* and *Sigma 2* were essential.

Strapped to her seat, Marina held a pencil out in front of her. More sensitive than any of their instruments, when it began a slow dip toward her chest, she knew they were just beginning to touch the atmosphere and to decelerate: inertia had kept the pencil traveling at the same speed while the spacecraft was slowing down. At their respective control panels, Vladimir and Sachi were peering intently at their instruments. The aerobraking maneuver was under computer control; their only function was as expert observers. If something went wrong they would take control and manually fly a timed profile into the atmosphere, using an artificial horizon and the g meter for reference. But that would be very crude by comparison, they admitted, and they wanted to save every drop of fuel to trim their orbits after the aeroshells had done their job. At .05 g, a panel light came on and the crew started timers. The rate of g onset was the only seat-of-the-pants method of estimating their entry angle. But it looked good, aboard both craft, as the g's slowly built. After eleven months their bodies were hypersensitive. One g, the acceleration they had spent their lives enduring, now seemed like four. Four g's felt like sixteen. Then the load slackened, and as they climbed back up out of the atmosphere, they returned smoothly to weightlessness. Their velocity had slowed sufficiently so that they were now captured by Mars's gravitational field. They had arrived.

CHAPTER

XXIII

MANGALA BASE

IT WAS hard to relax in Mars orbit. There was too much to see, to do, and beyond that a feeling of anticipation filled both spacecraft. Aeroshells were jettisoned and *Sigma 1* and *Sigma 2* used their rocket motors to trim their orbits into an ellipse whose low point (periapsis) was six hundred miles above the surface and whose high point (apoapsis) was eighteen thousand miles. This orbit, designed to conserve fuel, brought them over Mars's equator once a day. *Sigma 2* also released a small relay satellite to make communications more frequently available. Then the rendezvous process began and after a half-day chase *Sigma 1* closed the gap and docked with *Sigma 2* without incident. Next they turned their attention to the spectacle below them. As part of their training they had watched tapes of lunar orbits taken by the

Apollo astronauts. They had been disappointed in the monochromatic rock pile, an endless series of craters that all looked pretty much the same. By comparison they were overwhelmed by the variety of features below them. It wasn't the Earth, of course, but it was like a desert Earth, one devoid of water and vegetation. It looked, thought Wesley, like the stretch between El Paso and San Diego, if one ignored the missing highways and Salton Sea.

They were passing east to west and close to the equator. Dominating the view below was the great rift valley named after *Mariner*, the crude spacecraft that had first examined Mars nearly forty years earlier. Valles Marineris was thought to be like the Great Rift Valley of Africa, caused not by water flow but by faulting of the crust, with immense blocks of rock dropping between more or less parallel lines of subsurface weakness. Valles Marineris, if stretched across a map of the United States, would meander all the way from New York to California, with countless canyons and tributaries feeding into it, like bones attached to a fish's spine. It featured deep cracks and fissures with sheer drops of a mile or more. Immediately adjacent were flat-topped mesas that had been lifted up from below.

Thousands of channels fed into Valles Marineris and showed clear evidence of having been formed by the flow of water. Through their telescope the crew could see terraces that had been cut into the banks of these now dry arroyos by successive floods. The serpentine paths of these channels indicated they were the result of a flowing fluid rather than of crustal fractures. The absence of volcanic vents was a strong argument in favor of their having been created by the flow of water, not lava. Where had all that water gone?

The morning sun illuminated one end of Valles Marineris

but not the other. This caused differential heating of the carbon dioxide atmosphere, which in turn triggered a flow of wind through the canyon. They could see swirls of dust kicked up as the wind howled along, causing vertical reddish plumes to billow up above the canyon walls.

Valles Marineris was located in the eastern part of a region called Tharsis. As *Sigma* proceeded toward the west, the ground sloped up slightly toward the dome of Tharsis. At its deepest, the floor of Valles Marineris was twenty thousand feet below average ground level; at its peak Tharsis was twenty thousand feet above this datum plane. There were four volcanoes on the Tharsis dome: three in a row and then a larger, solitary one. Their alignment reminded Wesley of the head of the constellation Scorpius, whose brightest star, Antares, would correspond to Olympus Mons, the granddaddy of Martian mountains. Olympus Mons was three times the size of Mauna Loa, its nearest terrestrial competitor, and three times as high as Mount Everest. At its top was a depressed basin, or caldera, formed by lava cooling and slumping after the volcano's active phase had ended. There was some speculation, however, that a small amount of volcanic activity might remain, and as they passed overhead they took photographs and peered down into the caldera's nooks and crannies. The caldera was forty miles wide, and it was no easy job to map it. The results were inconclusive; they saw no smoke or steam rising, but several black spots could have been new lava emerging from small vents.

Their analysis was important because the caldera of Olympus Mons was, when they left Earth, their second choice for a landing site. Some scientists thought that if there were life on Mars it would be found in a place that provided minerals, water, heat, and protection from radiation. A volcanic vent in the shadows of the caldera wall might be exactly such a spot. Here life might still survive from the time when Mars was wetter and warmer and had a thicker atmosphere. Hot springs on Earth, in places like Yellowstone, were cited as examples. And deep in the oceans of Earth, life-forms had been discovered flourishing on a strange hydrogen sulfide metabolism, independent of sunlight. Vladimir was all for landing inside the caldera if he could find an appropriate spot, flat and smooth. The onboard file of Mariner, Viking, and Mars Observer photographs did not reveal enough detail to suit him, and he spent long hours looking through the telescope and taking photographs with a telephoto lens.

Wesley, on the other hand, favored their primary landing site. He was not as optimistic as Vladimir and Elena about the possibility of life on Mars, and as a geologist he was intrigued by the diversity of Mangala Vallis. In the eastern portion of this valley, ancient river channels, lava flows, and impact craters were clustered together. Also, there was an abundance of fine landing sites. In the end, after days of discussion with Mission Control, Mangala won out, primarily because it seemed safer and because no clear evidence of active vents had been found in the caldera—or anywhere else for that matter.

They were eager to land, but they also wanted to make absolutely certain that, prior to committing, they had examined the planet sufficiently and had once again rehearsed every phase of the landing sequence. If they were successful, Mangala would be not only *their* landing spot, but also the location of the first Mars base, for they would leave behind their supply craft with its nuclear power plant, their habitat structure, and the Rover. The next expedition would expand from these facilities. Their decision had long-range implications, perhaps for the remainder of the twenty-first century, and they owed it to those who would follow to establish the best possible beachhead. Mars had its bad-lands, jumbled and chaotic; it also had smooth, monotonous plains. The trick was to find a compromise between safety and science, a spot that would eventually grow into a self-sufficient settlement. Mangala seemed to fit the bill, and after two weeks of study, discussion, on-board rehearsals, and analysis by Earth-based experts, they were ready to give it a try.

Heather's arm was out of its sling now, but full strength had not returned. A large contingent of doctors, brown and fit after a week on the Black Sea coast, arrived at Mission Control with the news that the landing itself would not harm Heather, but that the arm was not up to the strenuous work for which she was scheduled during traverses in the Rover. Heather had to agree with them. Her arm seemed to have no strength at all. What the doctors kept to themselves was their discussion of the causes of Heather's accident. Was the impact she described sufficient to break a normal bone, or had weightlessness caused so much demineralization that the collarbone simply snapped as if she were an eighty-year-old with an advanced case of osteoporosis? Whatever the case, Heather and Elena would have to switch jobs on the surface, a prospect that distressed both of them. All that training! And they would have been so good at their regular duties! Now they weren't so sure.

When not peering through the telescope, Vladimir kept busy checking out *Condor*. After being dormant for eleven months, the landing craft showed no signs of wear, but its hydrogen and oxygen tanks had lost more of their contents by boil-off than expected and had to be replenished from *Sigma 2*'s supplies. Apparently the preflight thermal analysis had been faulty: *Condor*—wedged between *Sigma 2* and its aeroshell—had absorbed more heat than expected. *Sigma 1,* in turn, transferred fuel and oxidizer over to *Sigma 2* until the two were balanced again.

While Vladimir was working on *Condor,* Henri and Marina were preparing the supply ship for landing. It would precede *Condor,* and in many ways its task was more difficult, having no eyes with which to measure last-minute adjustments. Deciding which craft went first was linked to the rule that required supplies to be available on the surface before people could attempt to land. Also, to make sure the two vehicles ended up in the same spot, it would be easier for Vladimir to have an electronic beacon on the supply craft to home in on. His would be a simpler task than that faced by Pete Conrad, who flew *Apollo 12* to a precise landing next to the *Surveyor III* spacecraft, which had been resting for three years on Oceanus Procellarum, the Moon's Ocean of Storms.

There were two vital chores in preparing *Supply,* as they had named the craft, for landing. First the nuclear reactor aboard needed to be started up and tested, an intricate procedure even for an old hand like Henri. Once activated, the reactor would begin to emit harmful radiation, so Henri waited until as late as possible. However, it had to be done while still in orbit because the reactor's electricity was used to power *Supply* during descent and landing. The second job was to program *Supply*'s computer with precise instructions on how to reach the landing site. This was done by entering the geographical coordinates of Mangala into the computer, and then transferring to it a state vector—a mathematical depiction of location, velocity, and time—from *Sigma*. The state vector was divided into seven components: three indicating position in space, three for speed at that position, and one for the exact time at which the previous six were measured. As long as *Supply* remained attached to *Sigma*, their two state vectors were identical. When they separated, *Supply*'s instruments would sense changes in velocity due to rocket firing or atmospheric drag, and the state vector would be updated accordingly. Near the ground, radar information would be measured. The computer would then calculate steering commands to bring the state vector's position and velocity (relative to the landing site) smoothly to zero. That constituted a landing—with luck a gentle landing in the right spot.

Holding their breath, they released *Supply* and moved *Sigma* out of its way slightly. As programmed, *Supply*'s rocket engine burped briefly. They watched a tongue of flame shoot from it, and then *Supply* slowly receded behind them, losing altitude very slowly. By the time it unfurled its parachute, they were over the horizon. When *Supply*'s retrorocket cut in they were nearly on the other side of the planet. They could only wait. Their communications relay satellite was ahead of them in orbit, and as they came around the eastern limb of the planet they could hear a plaintive beep. It was *Supply*'s beacon, but coming from where? As Mangala swept over the horizon the weak, relayed signal was replaced by a strong direct beep and Vladimir flew to the telescope, nearly colliding with Marina. As usual, she had nothing on but her underwear, but for once he was not distracted. Anxiously he peered down at the now familiar features: the sequence of sinuous channels, a ridge line intersected by a massive fissure, with rubble at its base. Then just beyond, on the flat, ground zero. Finally he spotted good old *Supply,* somewhat to the northeast of where he thought it would be, but there it was, and it must be upright or its beacon wouldn't be chirping so merrily. There was much rejoicing over dinner that night.

Things were a lot different when *Condor* departed. There was a certain solemnity to the occasion, and the eight acted like people who met only occasionally in church. Henri and Marina came over from *Sigma 1* and joined Sachi and Mariko in *Sigma 2*. The four clung to the open hatch that led to *Condor,* and on the other side Vladimir, Elena, Wesley, and Heather gathered in a quiet little knot. Then they all seemed to speak at once, platitudes of no consequence. Sachi took photographs, and that was it. The last thing Vladimir saw before the hatch clunked shut was Marina's beautiful face. Then she was bathing in a mountain stream, and then he turned to *Condor*'s control panel. "State vector update," he barked, and Elena responded by opening her checklist with its long column of numbers. They had tracked *Supply* over three orbits, and their computer now knew their target's position to an accuracy of ten meters or so. For fifteen years the astrophysicists, mathematicians, and aeronautical engineers had argued about how best to bring them down on the surface with a minimum of fuel and a maximum of safety. The trip took just over an hour and involved a complicated sequence of events. First they fired a solid rocket motor mounted on the outside of *Condor*'s aeroshell. After a forty-five-second burn the motor was jettisoned, and they started a very gradual descent. Aerobraking began to become effective as they grazed the top of the atmosphere, and deceleration reached its maximum at an altitude of eighteen miles. At forty thousand feet the aeroshell was jettisoned, and at thirty thousand feet three parachutes were deployed. They slowed *Condor* from six hundred miles an hour to less than two hundred. When they were jettisoned at five thousand feet, *Condor* pitched over to a vertical position, and for the first time the crew could see where they were. Up to this point their approach angle had been such that, strapped in their hammocklike seats, all they could see was black sky. First the cockpit filled with an eerie pink glow, and their two windows were ablaze with the reflection from sun-drenched red rocks. They were looking at a canyon wall, an old one judging by its weathered appearance, with deep fissures and a bed of talus, or rock fragments, at its base. Their descent engine cut in, and they flew over the cliff at—according to their radar altimeter—six hundred feet. To Vladimir they seemed lower. Up ahead, his instruments told him, was *Supply* and its beacon, but all he could see was a plain littered with boulders the size of small automobiles. A dark lava flow intruded from his right, and they descended beyond what looked like a dried streambed. Finally a glint of metal up ahead: *Supply,* perched next to a small crater, seemingly undamaged. Vladimir took control from the computer and tested his skills briefly, rocking *Condor* left and right, up and down. Satisfied, he descended to about fifty feet and approached *Supply* gingerly: he could see from the shallow pit dug out by *Supply*'s own engine that the surface soil and rocks were loosely packed, and he didn't want his rocket exhaust to kick up rocks that might damage *Supply*. After making a complete circle of *Supply,* he could detect no wind. Noting with satisfaction that *Condor*'s descent stage still had 6 percent of its fuel remaining, he set the machine down—a gentle bump and an "engine off" acknowledgment from Elena, his copilot.

Behind them Wesley was nattering away about this and that rock formation, but Heather remained silent. Just in time, she thought; her bladder felt ready to burst. For several days now the landing party had been forcing down sodium-laced fluids to bring their blood volume closer to preflight levels and to reduce any tendency toward orthostatic hypotension and its accompanying light-headedness. Heather unstrapped quickly and stood up, then immediately sat down again. She had almost fainted. A moment later she tried again, gradually this time, and groped her way along the aft bulkhead to *Condor*'s tiny bathroom. One by one the others stood up and stretched. Gravity—even one-third of Earth's—felt strange and confining. They resented being pinned to the floor. Vladimir felt all right standing but the others were light-headed and weak in the knees. Their leg muscles and veins had forgotten how to cope with gravity. Their muscles were too lax and the flapper valves in their leg veins had neglected to close between heartbeats. As a result of this adaptation to weightlessness, blood was pooling in the lower parts of their bodies, and their brains were being shortchanged.

Their first task on the surface was to make sure that *Condor* was undamaged. Vladimir and Elena ran the subsystems through the prescribed checks and could find nothing wrong. Next, they had to let people know where they were and how they were doing. They would have liked to tell *Sigma* first, but it had disappeared over the western horizon and could not be reached for several hours, even through their relay satellite. So they called Mission Control. The eight of them had had several discussions about what to say but decided that a prepared statement was unnecessary: they would be spending forty days on the surface, reporting voluminously on their progress. Let the first words be spontaneous, not rehearsed. Elena controlled *Condor*'s radios—it fell to her. Somewhat self-consciously Elena cleared her throat and spoke very slowly. "Planet Earth, this is Mars calling, from Mangala Base. We Martians are fine, six percent fuel remaining, and *Supply* looks good." Vladimir and Wesley were applauding as Heather emerged from the bathroom. "What did she say?" "She said we had six percent fuel remaining," replied Wesley. "Why did she say a strange thing like that? That's not historic." Fourteen minutes later they received a reply: "Mangala Base, all the people on Earth rejoice with you in your historic accomplishment. Thanks to you, our universe is shrinking, becoming more habitable, and offers us limitless promise for the future. Your families send their love from your other planet, as do all twelve billion of us."

Their next task was to do nothing, and that proved difficult. Their adrenaline was pumping, and Wesley in particular couldn't wait to get over to *Supply*, release the Rover, and get moving. But they were well trained and they had agreed to spend as long as two days, if necessary, readapting to gravity. They needed to be strong and sure before donning pressure suits and going outside. Within an hour of landing all four of them felt fine, if somewhat tired. It had been a long and exciting day. As they relaxed, first they became hungry and then drowsy. *Condor* was not a great habitat—it was a small flying machine, and the four of them had to be satisfied with cold rations and cramped sleeping accommodations. The two smaller ones, Elena and Heather, slept in hammocks rigged from their lightweight fabric seats.

The two men slept on the floor. No one slept well in this strange new place. After about five hours Elena and Heather began whispering, and soon thereafter Vladimir got up and went to the bathroom. Only Wesley seemed comfortable, but the activity of the others soon roused him.

Elena conducted mini physical exams. When asked to stand on one leg and close their eyes, they could not keep their balance, not even for a second. On the aft bulkhead of *Condor*'s cockpit was a dart board identical to one aboard *Sigma* and another in their Houston training complex. Over the past year they had played the game enough to establish performance levels. After a short period of adaptation to weightlessness, their scores aboard *Sigma* soon matched, and then exceeded, their preflight averages. Now, in *Condor*, their scores were terrible, off by around 25 percent. They needed more time to reestablish muscular coordination and balance before getting down to work. They spent all that day resting, exercising, and planning their first excursion with Mission Control. Wesley scanned the horizon with his video camera, and from this data, plus what *Sigma* reported from overhead, Mission Control was able to pinpoint their location. Their first foray in the Rover was then plotted and transmitted to *Condor*'s teleprinter in the form of a map. With that in hand Wesley planned a time line of events. They would limit their first excursion to three hours and not venture out of sight of *Condor*. Once they had tested their equipment and become accustomed to working in this reduced gravity, they could extend themselves. First they had chores around camp, to unpack *Supply*, inflate its habitat, and find a remote location for the nuclear power plant.

By the next day their physical performance had improved sufficiently for them to get started. One by one Wesley,

Elena, and Vladimir donned their pressure suits. They were bulky and uncomfortable, worse even than the penguin suits. When sealed inside them, the explorers couldn't wipe the sweat out of their eyes or scratch their backs. Unfortunately, without them they would be dead within minutes of leaving *Condor*. Mars's thin carbon dioxide atmosphere was great for making dust storms but not for breathing. Finally, nearly forty-eight hours after they had landed, one of them stepped from *Condor*'s airlock out onto the surface of Mars. Their checklist, for no particular reason, called for Heather to emerge first, but because of her collarbone she remained inside at the radio console. In her first transmissions to Mission Control she referred to "we." "We are depressurizing the airlock now. . . . We show zero pressure. . . . We are stepping outside." The crew steadfastly refused to say who went first. "We flew by the checklist," they would say. "We all did it."

While Vladimir and Elena set about activating *Supply*, Wesley collected what was called the contingency sample. On the chance that some emergency might cut short their stay, they wanted at least to bring back a few rock samples for analysis. Wesley selected about a dozen, some of which he picked up and some of which he chipped off the boulders that were half buried in the landing zone. He looked in rock crevices for lichens or any other evidence of primitive life but found none.

Vladimir and Elena removed a panel from the side of *Supply* and pulled a lanyard, releasing an aluminum ramp. At its top was a compartment containing the Rover. Before it could be driven out, an elaborate procedure was required to activate the fuel cells that powered it. Once that was done, they took the Rover on a five-minute checkout and then parked it next to *Supply* again. Removing a second panel, they released another ramp, this one leading to their nuclear power plant. They used the Rover's winch and cable to drag it out of its stowage compartment. Then they pulled it about fifty yards away, up over a small rise and behind a large boulder. It was connected by cable to *Supply*. Tomorrow they would inflate the habitat module that was built into the opposite side of *Supply*. By this arrangement they would remain well shielded from the reactor's dangerous radiation. If subsequent expeditions wanted to expand their camp in the direction of the reactor, they could erect a protective berm of rock and soil around it.

That night they slept for the last time in *Condor*. Wesley dreamed of fossils. The next day was devoted to inflating their Mylar habitat and moving equipment into it from *Supply*. Then they deactivated *Condor* and transferred their belongings into the habitat. Mangala Base was operational.

The following day Wesley and Elena made their first short traverse in the Rover, bringing back a few rock samples and a lot of photographs. During the trip Wesley missed his wife and partner Heather, the Rover expert. He had never been sure exactly how a fuel cell worked, and now his life depended on a pair of them. Heather's presence would have reassured him in a way that Elena's never could. But as the days went by, their forays increased in distance and complexity. The Rover's inertial navigating system made it virtually impossible to get lost; backing it up was *Sigma* overhead, which could track them and establish their location. But if they could not get lost, they could get stuck. In this regard Elena was a lot more conservative than the reckless Heather would have been. Aboard the Rover they continually worried about going into places from which exit would be more difficult than entry. Principal among these were the arroyos, or dry riverbeds, that dotted the area. It was easy enough for the Rover to skid down a steep bank into one, but before doing so Wesley and Elena would, if necessary, detour for miles to make sure there was a safe slope to use when they wanted to climb out.

River channels were of special interest because Wesley and his Earth-bound cohorts thought that if ground water existed on Mars, it was apt to be found in such places, especially beneath the outer perimeter of bends or elbows. That was where it was apt to pool on Earth. Likewise, if life had existed at one time on Mars, fossilized traces of it might be found in the layers of rock and soil exposed along the steep banks, where water had once carved deeply into the surface. Like the walls of the Grand Canyon, the Martian riverbanks were a slice through time, a sequential map of the geologic history of the region, with the first chapters on the bottom and more recent developments in ascending order above. Wesley paid particular attention to rock crevices, searching for discoloration that might indicate colonies of bacteria. He also turned over plenty of small boulders in a similar search.

The complexity of their investigations was reflected in the equipment they carried. The petroleum industry had been a great help to them. Although they were prospecting for water, not oil, the techniques were similar. Their seismic equipment was slightly modified oil-industry gear: explosive charges, cable, geophones, seismometers, recorders. Their drilling equipment came not only from the petroleum industry but from the old Apollo Moon program. For boring down a few feet they used the lightweight, automated Apollo core samplers that operated independent of the Rover. For greater distances (and they drilled down several hundred feet) they used a more powerful drilling rig built into the Rover. Some core samples were sealed for analysis after return to Earth, but most were taken back to the habitat and examined there. (Elena thought it a bit strange that they had the ability to X-ray rocks but not people. Heather's collarbone, for example, might have mended better had *Sigma* been better equipped.) Heather prepared thin sections of rock for examination under a microscope and used a gas chromatograph and mass spectrometer to identify each element contained in the samples. She found that the surface samples, cooked by the Sun's ultraviolet bombardment, were much different chemically from their subsurface counterparts.

The work was designed not only to analyze conditions as they found them but to conduct experiments that would assist those who would follow. Moisture extracted from permafrost: Would it grow plants? How to purify it and produce drinking water? When and how could a colony produce enough water to sustain itself, or even enough surplus to expand? In addition to permafrost collection, could water be extracted from the atmosphere? How much energy was required to obtain oxygen from rocks? How about producing methane gas for fuel? In theory, water and carbon dioxide could be combined to form CH_4, methane, but was such a process practical here? How did Mars's reduced gravitational field affect manufacturing and agricultural processes? The questions seemed almost endless. As always seems to happen in science, each answer begot a new series of inquiries.

The time passed quickly; their forty days seemed woefully inadequate. Wesley and Elena became an effective team and eventually ventured more than a hundred miles from Mangala Base in overnight forays. The two balanced each other well. Wesley grew more impetuous. The field geologist in him always wanted to see what was over the next hill or inside the next crater. He wanted to drive and drive. The horizon, never more than two miles away, seemed too restrictive. He was falling in love with this planet but he was frustrated because he couldn't really feel it. He wanted to rip off his helmet and savor its eternal stillness without the hiss and rasp of his oxygen supply. He wanted to see its true colors without the modifying tint of his visor. He wanted to scoop up dirt in his hands and taste it. He wanted to be a part of this planet somehow. Elena, on the other hand, became increasingly cautious. She worried that they were overextending themselves, and she constantly checked their position in relation to Mangala. She interrupted their work to consider the latest news relayed to them by Mangala or *Sigma*. She wanted to get back home.

At the stopovers most distant from Mangala, Wesley and Elena left miniature weather stations. Their signals (temperature, pressure, wind velocity) were received aboard *Sigma* and rebroadcast to Earth, where they found their way regularly into the evening news. "Barometer dropping rapidly at Mangala North tonight, folks. Temperature minus eighty and wind out of the east at two hundred and twenty miles per hour. Tomorrow's forecast right after this announcement. . . ." Despite such extreme conditions, Mars's wind and temperature never stopped Wesley and Elena. Their pressure suits, like those of the Apollo Moon explorers before them, were so well insulated that they were almost insensitive to outside air temperature. The Martian atmospheric density was so low that a wind of two hundred miles per hour felt like a gentle zephyr back home. Sometimes dust storms on Mars obscure the entire planet for months, usually during summer in the southern hemisphere, but the explorers were lucky and never encountered anything worse than a mild restriction of visibility. The clarity of the atmosphere was most easily gauged by comparing the tint of the sky at various angles. Usually a diffused orange-pink band was visible just slightly above the horizon, while overhead the sky was dark, almost black. As more dust was blown into the atmosphere the pink expanded upward and the increased diffusion of sunlight caused a rosy glow.

Regardless of weather, the colors in the Martian countryside were spectacular. At one end of the spectrum, crystalline rock strata sparkled like diamonds in the sunshine. At the other extreme, black glass formed by lava flows was dull as asphalt. In between, the Red Planet lived up to its name. There was every variation of red, depending on the composition of the surface and the angle at which the sunlight struck it. Cliffs could be maroon at dawn and then lighten to crimson and blood red as the Sun rose. The Sun appeared 44 percent smaller from Mars than from Earth, but without Earth's thick atmosphere, Martian observers noted very little difference. It was still far too bright to stare at, even through a tinted visor, and its white light was a powerful yet subtle landscape painter. It brushed the hillsides with redness of every shade. The soil's rust-colored iron oxide was the visual point of departure. From it developed copper and orange, mustard and ocher, pink and rose, russet and burnt sienna.

Counterpoint to this variegated panorama was the placid sky, now pale, now dark. Through it drifted fragile wisps of clouds. Martian clouds are never dense, nor do they show much vertical development, as do Earth's cumulus stormclouds. They are somewhat like thin layers of stratus, but more closely resemble cirrus clouds. Sometimes they come in washboard waves, but usually they are more solitary. The *Sigma* crew often could find them over high-terrain features, such as near the peak of Olympus Mons, where they generally formed a partial circle. Viewed from above they tended to be milky white, while from below they were pastel pink or pale rose. But from whatever vantage point, the clouds were always delicate, an aerial filigree compared to Earth's thick, robust formations. Occasionally land and sky seemed to blend, as dust devils whipped the fine soil into whirling columns.

Phobos and Deimos added more variety to the sky. Deimos, the smaller, would come up in the east and move very slowly across the sky, taking two Mars days to reach the western horizon. While Deimos looked like a pinpoint of light, just slightly brighter than a star, Phobos, closer and larger, looked like a proper moon, with phases that waxed and waned. Phobos, rising in the west and setting about five hours later in the east, looked to our explorers to be less than half the size of Earth's moon. Its dark color made it appear drab; a full-moon night on Mars was darker than on Earth, not bright enough to read by.

During the landing party's forty days at Mangala Base, the quartet aboard *Sigma* was not nearly as busy as the four on the ground, but they still performed a wide variety of duties. Monitoring and maintaining the systems of the big machine took several hours each day. They continued their examination of the surface, looking for signs of active vulcanism (steaming vents or widening pools of lava), and they photographed anything that looked remotely out of the ordinary. Hundreds of pictures were taken of Phobos, even though they never came closer to it than five thousand miles. They measured their orbit with extreme precision, for minute variations indicated a less-than-uniform gravitational pull by the planet, revealing concentrations of mass below its surface. These mascons, first discovered during Apollo orbits of the Moon, could assist scientists in determining the history of Mars's formation and its inner composition. If extreme, mascons could also affect their rendezvous when *Condor* ascended. The crew of *Sigma* also spent a lot of time talking with Mission Control and occasionally acting as a communications link with Mangala. Toward the end of their stint they spent more time just waiting and worrying about their compatriots on the surface.

When the time came for *Condor* to leave, there were mixed emotions on the surface. Vladimir and Elena were anxious to get moving and started preparing for their departure several days earlier than necessary. Heather seemed preoccupied with her laboratory work, and Wesley was downright frustrated. How little they had accomplished, he thought, and how much more remained before Mangala Base could really be considered habitable for an extended stay. Maybe von Braun's original Mars scheme would have been better, he mused. Von Braun had proposed gigantic airplanes that would glide to a landing on one of Mars's polar ice caps, disgorging explorers who would trek over three thousand miles to the equatorial regions before departing. No, Wesley admitted, he knew that plan was impractical, but still—he had seen so little of this geologist's dream. Vladimir and

Elena bustled about with paper and pencil. Vladimir had a thousand ideas for improvements, for streamlining procedures, for new equipment that must be brought from Earth, for notes to leave here and there for their successors. Some instructions were quite detailed, while others—well: "This machine *will* work. Just kick it a couple of times." Heather's final act was to leave powdered milk and plastic-sealed cookies just inside the habitat's entrance. "Well, we used to do it every year for Santa Claus. Why should we do less for Martians? And besides—we don't have to worry about mice getting it." Vladimir pretended to be outraged at the untidy arrangement, saying that some of Heather's strange ways were the result of her childhood pampering of Saint Nicholas.

The levity continued right into the final countdown. Perhaps it was their way to relieve tension. Buzz Aldrin, about to make man's first ascent from the lunar surface, had told Houston, "Roger, I understand we are number one on the runway," the same terminology he would have used taking off from a busy airport. When *Condor*'s single ascent engine fired, they heard a loud pop as they separated from the descent stage and then they were airborne. The g level was quite low as they climbed vertically, and then it increased smoothly as they tipped over to a nearly horizontal trajectory. They swayed back and forth rhythmically as their steering rockets made corrections. Now they were upside down; they had to peer into the upper corners of their windows to see the reddish surface whiz by. The engine shut down on schedule, and they coasted until their radar found *Sigma* and they could make a few small corrections. With radar aboard both *Condor* and *Sigma* it was relatively easy to close the gap between the two. The *Condor* crew breathed a collective sigh of relief when they could actually see *Sigma* growing in their windows. They took their time, letting orbital mechanics rather than brute power do most of the work. Six hours after lift-off the docking latches snapped shut, and the eight men and women were reunited.

Once the four explorers were back inside *Sigma,* they moved *Condor* away by remote control and ignited its engine one final time. This was a retrofire burn to fuel depletion, and at its conclusion *Condor* began a death plunge back to the surface. When it crashed, the resulting Marsquake was recorded by the six seismic stations still operating on the planet.

The reunion began with yet another physical examination. For years the medical community had been debating what effect Mars's gravity might have on them after eleven months of weightlessness. The landing party had "calibrated" their bodies by treadmill runs before leaving Earth and again upon arriving at Mars. Now a third timed test showed their performance to be somewhere in between. Their cardiovascular systems were in better condition than they had been forty days earlier, but still they were significantly below their preflight norms. Partially rejuvenated by Mars's gravity, they still faced possible deterioration to dangerous levels during the nine-month trip home.

Once again *Sigma 1* and *Sigma 2* separated, this time for the engine burn that would break Mars's gravitational embrace and establish them on a slightly modified Hohmann transfer trajectory to Earth. Once again it was a time for worry. Having two round-trip modules added a lot of safety during coast phases of the mission, but a malfunction during an engine burn could render this redundancy totally useless. If, for example, *Sigma 1* made a successful burn and

Sigma 2 was unable to start its engines, *Sigma 1* did not have sufficient fuel to reverse course and go back and rescue the crew of *Sigma* 2. It was vital to keep the two craft on identical trajectories at all times. Therefore they separated only enough to maneuver without danger of hitting each other, and they ignited their engines at the same instant. They compared notes continuously during the maneuver (called the trans-Earth injection, or TEI, the same terminology used during the Apollo program). If one of them was to shut down prematurely during TEI, so must the other until the difficulty was sorted out.

Once safely established on their trans-Earth trajectory, the two craft docked for the last time and the crew settled in for the long ride home. No Venus this time, just their bodies and their machines against the clock.

CHAPTER

XXIV

HOME AGAIN

THE FIRST few weeks went by quickly as the crew relived their experiences and compared notes. For example, the southern hemisphere of Mars had seemed to be peppered with impact craters, while the northern regions tended to be covered by more recent lava flows. Why this celestial discrimination, to treat the halves of the planet differently? They had many theories, any one of which seemed as plausible as the others. Another on-board debate centered on the presence of water on Mars. Based on their core samples, there was plenty of subsurface water in a permafrost layer. Assuming Mangala was typical, there must be immense quantities of it locked in the soil. From orbit they saw clear evidence of vulcanism, past and perhaps present. Volcanoes meant interior heat, and heat would melt permafrost. Why hadn't they been able to find any traces of liquid water?

Communications during the present phase of the flight were constant and clear, so the crew invited Earth experts to join the discussion. A press conference for topics of general interest had already taken place, with the eight of them crowded around the video screen in the Yellow Room. It had gone on for hours, and worked so well that Wesley decided to organize a geology conference with Mars experts back on Earth, to get a head start on analyzing the data they had acquired. It was a husband-wife effort with Wesley interpreting the field geology while Heather reported on the chemical studies conducted in Mangala Base's laboratory. Preparing for the teleconference kept the two of them busy for more than a month, and the others pitched in and helped in whatever way they could.

Except Sachi. It became increasingly apparent that something was wrong with Sachi. He had always been quiet, but now he never spoke unless spoken to, and then gave short and occasionally nonsensical answers. Mariko was worried about him and discussed it with Elena, her fellow physician. Elena rarely saw Sachi and tended to think Mariko's fears were exaggerated. But Vladimir took them seriously and began to make special efforts to entice Sachi into group activities. The sims, for example, continued periodically at Vladimir's insistence, although it was clear that the crew's appetite for them had largely evaporated. During sims Vladimir made his best effort to draw Sachi into the troubleshooting discussions. Occasionally it worked, but mostly Sachi balked, saying simply that the flower told him everything was all right. The flower, according to Mariko, was the artificial yellow rose that someone had taped onto the main console in the Yellow Room. There Sachi spent most of his time, because the flower's voice, like a speaker wired into his own private Mission Control, told him of things on Earth.

The CELSS in *Sigma 2* got worse and worse. Not only were the toilet and the shower unusable, but now a rank stench filled the compartment. They had tried swabbing various disinfectants on the walls (they had no deodorants on board), but nothing seemed to help. Finally, in frustration, they sealed off the bathroom except for periodic inspections. To their horror they discovered a greenish yellow crust on the surface of the toilet and throughout the shower stall. Clearly it was spreading. Mission Control commiserated with them but was not helpful in offering solutions. Scraping the gunk with a kitchen knife seemed to work better than anything else, but then—one by one—the scrapers developed persistent colds, with coughs and mild fevers. Next *Sigma 2*'s water supply turned sour, and a plastic tube of it was no longer clear but tinged with brown. Those who drank it got diarrhea, and Vladimir ordered the supply valves turned off. He also forbade any showers for the time being, and everyone used *Sigma 1* as a kitchen and bathroom. Those who lived in *Sigma 1* never left it, except for Marina, who was supposed to be the CELSS expert and felt an obligation to visit and inspect *Sigma 2*.

One day Marina floated over in her underwear and Sachi attacked her, more or less. It never was clear whether his intent was sexual, or if he simply wanted to share more intimately with her the latest news from the flower. At any rate he grabbed her, and Marina screamed. That brought the others, and Sachi did not react well. He retreated to his sleep chamber, yelling in Japanese, a language he almost never used. Mariko went after him but emerged minutes later, weeping. She could make no more sense of what he said than could any of the others. The crew of *Sigma* was now only seven strong, with one alien among them. Vladimir felt that as commander he needed to do something about Sachi, but he didn't know what. There wasn't enough sedative on board to last for months. Kill him if he got violent, Vladimir thought. He didn't want to do it, but what was a commander for if not to make hard choices? They had no weapons on board, but still there were ways. Probably a hypodermic first, then out through the airlock. He should be prepared. As a precaution he ordered Marina to stay in *Sigma 1* and Sachi not to leave *Sigma 2*. Both orders were superfluous, he knew, because Sachi wouldn't leave his flower and Marina would not risk another encounter, but he felt better for having taken some action.

In the meantime at least the clock was running and they were getting closer to home. The only device more popular than the clock was the calendar. Wesley and Heather marked the passage of each twenty-four hours with a little ceremony and another bold X. The seven of them celebrated their second space Christmas in *Sigma 1,* leaving Sachi alone in the Yellow Room. Even with a little medicinal brandy it was a melancholy occasion, but despite Sachi, their colds, and the CELSS problem, the group felt more optimistic than they had a year earlier. They certainly were more sure of themselves, more able to cope. They were veterans, not rookies.

Mariko took throat cultures from all eight and examined them under a microscope. The four who lived in *Sigma 1* showed normal microflora, but the *Sigma 2* contingent did not. In addition to the usual bacteria, she identified several shapes that were unfamiliar to her and that she could not find in her small reference library. Similar shapes were in the algaelike growth in *Sigma 2*'s shower stall. She had no way of transmitting these tiny images to Earth but she did describe them to Mission Control and received the usual reply: "Stand by." A hastily gathered meeting of the medical committee (in Key West) yielded a tentative diagnosis of "genetic mutation of microflora due to environmental factors." "What's that mean?" Vladimir wanted to know, "the environment of space or the environment of Mars?" "Nobody knows," replied Mariko, "but probably Mars has nothing to do with it. Our CELSS was acting up before we got to Mars, and anyway, weightlessness plus radiation is sufficient cause of something like this. What it means, though, is they are going to want to keep us in quarantine until they understand it better. How long? Who knows?"

While Sachi had his flower, the others had theirs too. For most it was watching television. Their library of movies and games was extensive. Radio chats with people on Earth were also popular. Celebrities from a variety of fields were delighted to be invited to share some time with the crew: musicians, sports figures, politicians, scientists—all had their egos tickled by being included in a mission that was played out on the world stage. Cocooning also became popular with some. Henri and Marina now spent more and more time in their sleep compartment. When teased about it, they admitted their sex had never been better. "Ah, weightlessness"—Henri rolled his eyes—"I only wish we could have tried it on Mars." A general feeling of lassitude pervaded *Sigma,* a feeling Vladimir tried to combat. He rushed about—checking, inspecting, cajoling, urging, coercing. In the process he became less and less popular, and at sim time (which now excluded Sachi) there were several near revolts.

In addition to sims, Vladimir pushed exercise. They all knew he was right, but the two hours a day (with only sponge baths instead of showers) became increasingly difficult. Food was also a problem. Each of them had eaten more than fifteen hundred meals on board, and by now everything tasted bland and dull. Some genius on their loading crew had filled a compartment labeled as fruit cocktail with pickled beets. Fruit cocktail was one of their few favorites, and they had been hoarding it. Now that hoard turned out to be the despicable beets. Little things loomed large, and tempers were frayed. With two months remaining, Vladimir was afraid he was losing control.

Then one day in February his own world disintegrated. Mission Control notified him and Elena, after much hemming and hawing, that their son had been killed in an automobile accident near Moscow. The pair reacted by retreating to their sleep compartment. The others were able to coax Elena back out, to resume the trappings of a normal life—if life aboard *Sigma* could be called normal—but Vladimir remained cloistered except for necessary forays. He stopped shaving. No more sims, no more inspections. He seemed to have resigned as commander. Oddly enough the news had a salutary effect on Sachi. The flower frequently brought him death announcements and now, in one instance at least, a report had been corroborated. Also, with Vladimir not to be seen, Sachi felt he was needed to keep the equipment operating. He emerged partially from his shell and although he still spoke irrationally at times, it was now possible to get useful work from him.

By early March the Earth had changed from bright star to planet, through the telescope at least. Once again they had something to look at, to give them some measure of their progress. Elena coaxed Vladimir into taking a peek, and he slowly began to return to his duties. When the time came to activate *Peanut,* their Earth-capture vehicle, they all pitched in with a vigor absent the past six months. When *Peanut* passed its tests, Wesley for the first time allowed himself to think that they really were going to carry off this entire improbable venture. He even had a radio conversation with his stockbroker and did some investing. Counting his and Heather's modest but unspent salaries, twenty-two months' worth, they had quite a nest egg. It was time, he thought, to reacquire a taste for Earth and its delights.

When the time came to enter *Peanut* and cast loose from *Sigma,* the eight felt varying proportions of sadness and elation. *Sigma,* their mostly happy home for the past twenty-two months, would continue on, swinging past the Earth at more than twenty-seven thousand miles per hour in its elliptical orbit around the Sun. As they passed within the orbit of the Moon the crew saw what the Apollo astronauts had described, an Earth that was a blue and white pea that could be covered by one extended finger. Three times the eight had seen an approaching planet, one yellow, one red, one blue—the first humans to do so. Compared with Mars, the Earth was a gleaming headlight, its oceans reflecting sunlight back into their eyes more efficiently than any Martian surface. Blue oceans and white clouds dominated. They were hard pressed to identify land masses except for the rust-colored Atlas Mountains of North Africa.

They were headed for space station *Freedom,* which in their absence had been converted into a quarantine facility.

Before they could dock with it, however, they had to slow from *Sigma*'s supercircular speed to *Freedom*'s orbital velocity. They used Earth's atmosphere to decelerate, plunging into it aeroshell first at a higher speed by several thousand miles per hour than humans had ever traveled before. Tracking radars had been following their trajectory for days, and only a minute course correction had been needed aboard *Peanut*. Vladimir noted with satisfaction that the small craft maneuvered with the agility of a fighter plane. He was flying from the left seat with Sachi on his right as copilot and the others behind. Since neither Vladimir nor Sachi was exactly 100 percent, the others fidgeted and looked over their shoulders, but none of them was trained to take over, and things were happening too fast anyway. First they grazed the atmosphere at a shallow angle and then dived more steeply. One-two-three-four-five g's: the unaccustomed pull on their bodies pinned them in place. Then the worst was over, and the g's abated as they began to climb back up out of the atmosphere, this time at subcircular speed, below the threshold at which Earth's gravitational field would retain them. From there they relied on rocket power to make the necessary orbital adjustments to put them in the same plane with *Freedom,* below and behind it. Then they caught up gradually and docked gently.

Aboard *Freedom* a special quarantine crew of four awaited them: a commander (also pilot and systems engineer), a biologist, a chemist, and a physician. The station was crowded with the dozen of them, and they didn't know how long they would have to stay. Nonetheless they were overjoyed to have transferred into different surroundings, to find new faces and voices intermingled with theirs. As usual each new phase of their mission required a physical exam.

The four who had lived aboard *Sigma 2* were now over their colds and were as healthy as the other four. Blood and urine samples were taken, as well as throat cultures. A scanning electron microscope produced some strange new sights, but all the pathogens in their bodies, whether of terrestrial origin or not, were vulnerable to penicillin and other antibiotics. Their rock and soil samples passed an extensive battery of tests. Colonies of bacteria were grown in cultures and then destroyed by radiation or medication. While these tests were being conducted, the eight were given flu shots and other oral prescriptions designed to protect them during their reintroduction to terrestrial contagion. They also began to exercise strenuously, and to drink saline fluids as preparation for the surface and its strong gravity field. Perhaps because of guilty consciences, the treadmill and bicycle were kept whirring almost around the clock. Finally, after two weeks the microbiology committee (in Palm Springs this time) agreed that Earth would not be harmed by their return. It took two shuttles to deliver them and their cargo to Edwards Air Force Base, California.

Returning to Earth was almost as much of a shock as leaving it. Elena had the mildest reaction to gravity, Henri the most severe. Elena was up and walking without assistance within minutes. Henri had to be taken off the venerable *Columbia* in a stretcher and spent the night in the base hospital, being given the VIP treatment. The next morning, April 11, 2005, all eight resumed their lives as Earthlings.

CHAPTER

XXV

COLONY

PLATO IS said to have considered the sphere the perfect form. That may be, but peering into a crystal ball is also looking through a spherical lens. Distortion is inevitable. My perception of the first expedition to Mars is just a keel and a compass. I have tried to follow the Law of Least Astonishment—that is, to be conservative in judging how the mission will be conducted and what it might discover. The only part that is not conservative is the timetable. The year 2004 is approaching fast, perhaps too fast, and I have picked it as launch date partially out of frustration at today's slow pace of events. The flight *could* take place just as I have described, but it almost certainly will not. Better ideas will come along before the twentieth century ends. Advances in propulsion may make the chemical rocket as obsolete for deep-space applications as the slide rule is for math. Fascinating—and entirely possible—is the notion of arranging collisions between protons and antiprotons, thereby releasing intense bursts of energy. Such antimatter engines may come into being by century's end. I have stressed in various ways (including the mess in *Sigma 2*'s bathroom) the progress that needs to be made in CELSS research. I don't think dramatic breakthroughs are in the cards here, just a lot of hard work. But before that work will begin in earnest, NASA needs to recognize the problem and give it a much higher priority—and that means more money.

I hope that I have also made clear that the first landing, remarkable as it will be, should not be considered an end in itself. On the contrary it will be the "one small step" that leads to a self-sustaining colony. Of our eight pioneers, most will disappear into the folds and crevices of the good Earth, but I like to think that at least one will return to Mangala. Probably Heather. She will divorce Wesley, who has grown fat and indolent on the banquet circuit, and join the expedition of 2012. When she leaves she won't know how long she will be gone. Her first flight will have taught her patience, and she will be a good settler. Other specialists will accompany her, especially biologists and agronomists. Oxygen, water, food: those are the three primary consumables that Mars colonists must produce.

First water. There are three sources: the permafrost layer, the polar ice caps, and the atmosphere. There may be a fourth, reservoirs of underground ice or water, but certainly there are three. It will not be easy to mine permafrost. On Earth permafrost is extremely hard and tends to break drill bits. It won't be any easier on Mars. Once permafrost is excavated, it must be heated to yield liquid water. In all likelihood this water will not be potable, but will be a brine that must be purified. The polar ice caps will be much easier to work with than permafrost, but their location makes them inaccessible from a base near the Martian equator. Equatorial locations are preferred, at least initially, because from them it is easier to ascend to orbit.

That leads us to the third source of water. The Martian atmosphere is composed of 95 percent carbon dioxide, 2.7 percent nitrogen, 1.6 percent argon, and *a trace of water vapor*—.03 to .1 percent. Water can be extracted from the air simply by compressing and cooling it. All that is needed is electricity to power a compressor and a refrigerator. Some calculations indicate that it would require the power output of a small nuclear reactor (100 kilowatts) for about two and one-half hours to produce one gallon of water. Not bad. Solar power might be used in lieu of nuclear, but dust storms can impede the functioning of solar panels, not to mention that—at an average distance of 140 million miles from the Sun—a huge array of panels would be required.

Once water starts trickling in, the colonists' situation will be much eased. By passing an electric current through water, breathing oxygen can be obtained. Water can be used inside greenhouses to irrigate plants growing in a carbon dioxide atmosphere. The hydrogen in water can be combined with the carbon in carbon dioxide to produce methane fuel (CH_4). Methane and oxygen burned in a rocket motor produce a specific impulse of 340 seconds, which is low by Earth standards but acceptable for a locally produced propellant. This means that as the colony gains experience in manufacturing, rockets arriving from Earth could bring more cargo and less fuel, relying on Mars-produced methane and oxygen for the return trip.

The Martian soil consists primarily of oxides of silica, iron, sulfur, magnesium, aluminum, calcium, and titanium—roughly in that order. It is not known whether there is enough phosphorus to permit healthy reproduction of plants and animals, but the other constituents seem well suited for agriculture. On Earth, plants have flourished in lunar soil brought back by Apollo astronauts, and Martian soil should be equally fertile. It will be relatively easy to erect inflatable greenhouses, pressurized to some extent. Carbon dioxide plus sunlight plus water: the first two necessities are readily available. Colonists will have to supply the water and perhaps some phosphorus and other fertilizers. Human waste will certainly be used for fertilizer as it is in many countries on Earth. A by-product of growing plants will be oxygen, and there may be practical ways of extracting it from greenhouses. The first colonists will be vegetarians, or at least will rely on freeze-dried meat brought from Earth, but gradually that too will change. Chickens and rabbits both seem likely candidates, and perhaps goats for milk. I would not like to contemplate transporting a horse or cow to Mars.

The Martian soil must also provide shelter. The atmosphere is thick enough to incinerate meteorites weighing up to a pound or so, but it is too thin to protect animal life from ultraviolet radiation. To shield themselves, the new Martians will have to pile about a yard of dirt on their roofs or burrow into the soil. For millennia on Earth, caves have served humans well, and they will also on Mars. One of the reasons Mangala was selected was because there are canyon walls there that can be dug out for habitats. Inside a cave the temperature will be lower and more stable than outside, wind will be nonexistent, and the problem of pressurization will be eased. Later the Martian soil will be mined to produce cement, glass, and metals and more elaborate structures can be built. Cement is composed of the silicates and aluminates of calcium. All three of these elements are abundant on Mars. To produce cement, soil needs to be heated, and then more water is required to mix with the soil plus sand to form concrete. Again the availability of water will be the key. The hydrogen in water can also be used, in combination with carbon, to produce plastics and polymers.

As the colony grows, its inhabitants, like pioneers anywhere, will want to know what is over the next hill, around the next bend. Mobility will be a big problem. Unless inside a vehicle, travelers will have to wear pressure suits. Rovers, perhaps methane powered, will grow larger and capable of greater range. Small, remotely controlled airplanes (unmanned) may be used for reconnaissance, as may balloons or dirigibles, manned or unmanned. The polar ice caps will be destinations that Martians covet, as Moslems dream of a trip to Mecca. Eventually some will make it there (the North Pole is preferable because it contains more water ice and less frozen carbon dioxide than the South Pole). The explorers will take deep core samples and from them will read a partial history of Mars, just as scientists on Earth have done by boring through the ice of Greenland and Antarctica. Martians may establish a polar outpost and they may figure out a way to get water from the poles to equatorial regions, perhaps through a pipeline, as oil is transported on Earth.

In some ways they will wish Mars were more Earth-like, especially in regard to the atmosphere. Living under a dome will be restrictive, as the Soviets have learned in Siberia, but on Mars it will be essential, perhaps forever. The alternative is something called *terraforming*—making the entire atmosphere of the planet breathable by generating gases with a suitable combination of temperature, pressure, and chemical composition. Heating is the key to terraforming, and various schemes have been proposed to raise the temperature of the planet. For example, if the polar ice caps were black instead of white they would absorb more sunlight and would eventually melt. Released water vapor and carbon dioxide would thicken the atmosphere, which in turn would trap more heat from the Sun, and Mars's temperature would slowly rise.

Some have suggested using Phobos as raw material for blackening the surface of Mars. Phobos is very dark in color; pulverized dust from it, shot down onto Mars, especially onto the poles, would gradually start the temperature rising. Other ideas for heating Mars involve the generation of greenhouse gases such as Freon, or placing huge mirrors on Phobos to beam sunlight to the poles.

Even after the atmosphere became warm and thick, it would still be poisonous to humans. The preponderance of carbon dioxide must be replaced by oxygen and probably nitrogen. On Earth, some 3.5 billion years ago, algae and lichens began the process of converting primitive gases (ammonia, methane, hydrogen sulfide) into oxygen. They and other plants could be used on Mars. So might bacteria that could feed off the chemicals in the Martian soil. Some bacteria excrete oxygen, others nitrogen—both good for humans. Plants and bacteria would have to be protected against ultraviolet radiation, at least in the early stages of terraforming. That would be less of a problem for the bacteria, which can burrow underground. Some bacteria also reproduce as often as every twenty minutes. As the atmosphere improved, more complex plant forms could be introduced, and animals as well. Bioengineering is an emerging discipline today that might prove invaluable to Martian settlers. Certainly a resistance to radiation would be a desirable characteristic to develop in Martian flora and fauna. Likewise an appetite for carbon dioxide at very low atmospheric pressure. Genetic mutants that scientists would be reluctant to release on Earth could be tried on Mars. Biological research might become a major industry for a Mars colony, with the most successful new life-forms becoming candidates for export to Earth. Certainly the gardening of Mars would be vital for its long-term prosperity.

Another idea involves the massive importation—by collision—of raw materials. Phobos is a modest candidate. The orbit of Phobos is decaying and the tiny moon sooner or later will crash into Mars. On a cosmic scale, it wouldn't take much of a nudge to accelerate the process. But water from Phobos, although valuable, would not be sufficient to terraform Mars. Comets, asteroids, moons of Jupiter or Saturn: all have been proposed as megasources of water and other raw materials. The problem is how to alter the trajectory of one of these bodies enough so that it collides with Mars. Usually hydrogen bomb explosions are suggested. If a big enough projectile hit Mars, some experts believe, the impact would release interior heat and activate volcanoes. This activity, independent of the material added by the interloper, would trigger or accelerate the terraforming process.

By whatever methods, the terraforming of Mars will be an extraordinarily slow process by the standards of terrestrial civilization. A NASA scientist, Christopher P. McKay, breaks it down into Phase 1, warming the planet, and Phase 2, converting atmospheric carbon dioxide and soil nitrates to the desired mixture of oxygen and nitrogen. He estimates two hundred years for Phase 1 and a hundred thousand years for Phase 2. Of course, humans can live successfully under domes in the meantime, but still . . . On the other hand, human creativity working in a new place with new raw materials might accelerate the process by means we cannot today imagine. Certainly Martians will have an acute desire to improve themselves and their surroundings. Just as the first crew felt elevated by the attention of its world audience, and honored to be representing the home planet, so will early settlers have the strongest motivation to succeed, as individuals and as members of a growing community.

Similarities between Mars and Earth will be comforting to colonists. As on Earth, weather will be a big topic of conversation. There will be no rain, floods, or lightning; no snow, but a trace of frost and ice. Dust storms will be the big news. Most subside within a couple of days, but when *Mariner 9* arrived at Mars in 1971 it found a planetwide obscuration of the atmosphere that lasted well over a month. Martian dust, fine as smoke, will be very corrosive, and filtering it out will be a problem. Temperature extremes on Mars will be duly noted as they are on Earth. At a place like Mangala it may get as warm as 50 degrees Fahrenheit and as cold as 150 below. Always insulated and shielded from ultraviolet rays, the settlers will not experience these extremes as we do on Earth. Day and night on Mars will be remarkably Earth-like since the Martian day is only thirty-seven minutes longer than ours. Martians will see two diminutive moons in their night sky instead of our larger one.

The Mars Chamber of Commerce will boast of one quality unmatched—and unmatchable—by Earth: reduced gravity. At 38 percent of their Earth weight, settlers will revel in their newfound physical freedom of movement. A fifteen-foot high jump, a fifty-foot broad jump? No problem, and you don't have to be a trained athlete to do them. Almost every record in the terrestrial book will have to be annotated "record for Earth only." Ballet on Mars will offer possibilities far beyond terrestrial choreography. Dancers will stay off the ground much longer and while airborne will be able to perform with a languid grace and depth of expression not possible with abbreviated terrestrial leaps.

The wonder of the Martian gravity, like any new sensation, will fade with time. When the first child is born, a child who has never known Earth's gravity, a new era will begin. Martians will no longer have to import people, although they will continue to do so for breeding stock and for acquisition of expertise. For those who wish to leave, returning home will not be inordinately difficult because the demand for Earth's goods—at first necessities and later luxuries—means supply ships will be returning to Earth with room for passengers. And new techniques, such as drug-induced hibernation, may make the trip seem much shorter. But little by little, Mars will become independent and self-sufficient, psychologically as well as economically. A child born and raised there, in a biologically isolated environment, may be too fragile to endure a visit to filthy, bone-crushing Earth.

In the early years, a Mars settlement will be a more dangerous place even than Earth. Settlers will have a constant fight against the hostile environment—radiation, poisonous gases, low pressure, and temperature. But they will have several advantages not usually available to settlers. They will not have to face the enmity of aborigines they displace. There are no snakes or mosquitoes on Mars. At first, at least, the only diseases will be those imported from Earth. Perhaps an even greater advantage will be the ability of a Martian civilization to evolve from one tiny group that shares common goals. There will be no national borders with which to contend, no traditional enemies, no weapons—unless they are brought from Earth. The first settlers will not be all alike, but I believe the unifying influence of the long trip out and the daily struggle against the environment will outweigh terrestrial ethnic, cultural, and national differences. Martians will develop and nurture their own traditions and governmental forms. The election of the first mayor of Mangala may be a more important milestone than the birth of the first child.

If some tragedy should befall Earth, Mars will still be there. Partially on the basis of analysis of Apollo's Moon rocks, scientists now believe that the Moon most likely was formed when a huge object—about the size of Mars—collided with Earth. This wanderer from deep space struck the Earth early in its formation and with such force that it knocked off a huge chunk. The Moon congealed from the solid and gaseous remnants of this collision. It is doubtful that human beings could survive on Earth if a second such collision occurred. The chances of that happening are almost infinitesimal, but the consequences are almost infinitely large. To me, this possibility is not sufficient justification for going to Mars, but a colony there would preserve the human race in the event of its death on Earth.

Not long ago I spent a week celebrating—if that is the right word—the twentieth anniversary of the first lunar landing by Neil Armstrong and Buzz Aldrin. I think the three of us agree that for a variety of reasons the space program has bogged down since 1969, and that it is time to get moving again. In this book I have outlined why I think a Mars mission makes sense and how to get there. If the first landing comes early in the twenty-first century, it will be possible to have a small but flourishing settlement by July 20, 2019—the fiftieth anniversary of the lunar landing. By that time Armstrong, Aldrin, and I will be nearly ninety and probably surveying nursing homes. Perhaps on Mars. . . .

But this timing will not be possible if we decide that necessary preliminaries include building an elaborate space infrastructure, including a Moon base. To borrow a pet phrase of the physicist Freeman Dyson, "Quick is beautiful." If we want to go to Mars, we should go—simply and directly, without the time-consuming and expensive detours that supposedly would add safety to the enterprise. In my experience making things unnecessarily complicated usually detracts from safety.

Perhaps I am looking at Mars through rose-colored glasses (not an inappropriate color for that planet). Perhaps I am even one of those dreamy-faced loons. I am reminded of Lord Brougham's assessment in 1845 of an earlier generation of explorers: "They cannot help it, these arctic fellows.

It is in the blood." I think I had it in my blood before I flew to the Moon, but at any rate that trip whetted my appetite for a far finer planet, one midway between the verdant garden of Earth and the sterile rock pile we call the Moon.

When we do go, I hope the real travelers share the good fortune of the *Sigma* crew. Our Martians should emulate "these arctic fellows"—their guts and perseverance, their meticulous preparation. May their intrepid spirit fly to Mars along with that of Gagarin and Armstrong. What will the first explorers find? Is there life on Mars? Maybe not, but there will be.

INDEX

Q

R

T

U

V

W